INTERNATIONAL SERIES OF MONOGRAPHS ON
PURE AND APPLIED BIOLOGY

Division: **PLANT PHYSIOLOGY**

GENERAL EDITORS: P. F. WAREING and A. W. GALSTON

VOLUME 2

THE PLANT CELL WALL

To Bobbie, Art, and James

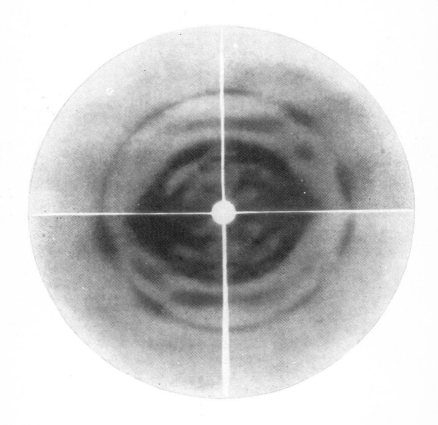

X-ray diagram of hemp fibres before and after delignification.
Upper right and lower left: "Untreated hemp fibres".
Upper left and lover right: "Fibres after lignin extraction".

THE PLANT CELL WALL

A Topical Study of Architecture,
Dynamics, Comparative Chemistry
and Technology in a Biological System

by

S. M. SIEGEL

Group Leader in Physical Biochemistry

Union Carbide Research Institute

Eastview New York

A Pergamon Press Book

THE MACMILLAN COMPANY
NEW YORK

PERGAMON PRESS INC.
122 East 55th Street, New York 22, N.Y.
1404 New York Avenue, N. W,
Washington 5, D.C.

PERGAMON PRESS LTD.
Headington Hill Hall, Oxford
4 & 5 Fitzroy Square, London W.1

PERGAMON PRESS S.A.R.L.
24 Rue des Écoles, Paris V^e

PERGAMON PRESS G.m.b.H.
Kaiserstrasse 75, Frankfurt am Main

Library of Congress Card Number 62–8705

Printed in Poland to the order of PWN-Polish Scientific Publishers
by Wrocławska Drukarnia Naukowa, Wrocław

CONTENTS

PREFACE

THE recent great advances in quantitative biology have been made possible by the growth of new concepts and techniques in chemistry, physics, and geology. These fundamental areas, together with medicine, agriculture, and technology have provided the stimulus for the modern inter-disciplinary approach to structure and function in living matter.

Among the more complex biological products, cell walls and intercellular substances are nearly unique in having been subjected to virtually all of the potentially applicable methods of study and analysis. It is indeed a measure of their complexity that we still know so little about these materials, although they have been objects for immunological, geological, colloid, and textile research. Nevertheless, endeavors in these and many other areas have revealed the outlines of a structural–functional system dynamically associated with the living protoplast yet possessed of considerable continuity in the fossil record.

If it were important for no other reason, the study of cell walls and kindred structures would provide an object lesson in the effectiveness of the broad approach. Of course, cell wall research is not of interest solely as an exercise in the unification of scientific methods and ideas. The macromolecular—and often highly structured–products which are manufactured at the protoplastic surface must reflect the interaction between the biochemistry of the cell interior and the physicochemical conditions at the cell boundary. In the constituents of the wall and their arrangement must reside a great body of information about such interactions and the state of the protoplastic surface.

No less important is the contribution which cell wall research can make to the study of evolutionary processes. It is most fortunate that there exist for study comparatively resistant substances such as lignin in plants and bone in animals, so that the proper combination of paleobiochemistry and comparative biochemistry, can provide new insight into the physiological and biochemical aspects of evolution.

This volume represents an effort to bring together some of this widespread and wide-ranging information, and to a modest extent, to make some attempt at the organization and synthesis of various facts and theories.

Although the results of recent and current research, including the author's studies on lignins and lignification, have been incorporated, this volume is not presented primarily as an advanced research treatise. The future of cell wall research, both in its own right and as an example of a valuable orientation in experimental biology, rests with the present-day students of botany and zoology who are willing to acquire understanding and proficiency in many areas of science and as much with those in chemistry and physics who have latterly become aware of the rich field of investigation that is biology.

To those who made this book possible, I take pleasure in expressing the deepest gratitude. Thanks are due particularly to Professor James Bonner, California Institute of Technology; Professor Arthur W. Galston, Yale University; and Professor David Goddard, University of Pennsylvania; who each in his own way contributed support, encouragement, and stimulating ideas to the author's research on lignification.

Thanks are also due to the students whose loyalty, help and enthusiam lightened many tasks and broadened many ideas: Mr. L. N. Chessin, New York University School of Medicine; Mr. K. Cost, University of Minnesota; Mr. N. Goodman, Brandeis University; Mr. B. LeFevre, Armour and Co., Chicago; Dr. D. Ridgeway, California Institute of Technology; and Mrs. Colin Taylor, University of Rochester.

To his wife and colleague, Barbara Siegel, the author expresses his indebtedness for many years of challenging ideas and constructive criticism.

Material, and in no small degree, moral, support was provided by the Wallace C. and Clara A. Abbot Memorial Fund of the University of Chicago; by the National Cancer Institute, National Institutes of Health, United States Public Health Service (Research

Grant C-2730); and by the John Simon Guggenheim Memorial Foundation.

Finally, the author is deeply indebted to the Union Carbide Research Institute and the Union Carbide Corporation for their generosity in seeing the manuscript through its gestation period, and to Mrs. J. Hess, Miss M. Wood, Mrs. F. Bouchard, Mrs. B. Clark and Miss G. Plock for their patience and cheerful efforts that guaranteed its safe delivery.

If there are controversial ideas in this monograph, the author is pleased to assume in full the responsibility for their presence and vigorous exposition.

INTRODUCTION: WALLS AND CELLS

WHEN the cell theory and its origins are presented to the student of botany or general biology, reference is almost invariably made to the cellular structure of cork as described in Robert Hooke's *Micrographia*. The existence of well-defined, limiting wall structures has been of great importance in the development of the concepts of biological organization. In the vascular plant body, a multitude of varied patterns of cell and tissue organization underlie a comparatively simple array of organs. The identity of tissues and tissue systems is based upon the cell types present and their arrangement, and is referred in large measure to cell morphology. In many cell types, form becomes fixed during maturation and differentiation in the cell wall. Thus, the death of protoplasts often leaves behind a permanent record of their size, form and arrangement. The stability of the morphological system in vascular plants is illustrated by the cellular and tissue structure which is often discernible in fossils.

Underlying the record of microstructure which the cell wall provides is a vast supply of physicochemical information encompassing the physical structure of polymer aggregates and polymer chains, and the chemistry of polymers, monomers and other small molecules. Therefore, a vital part of cellular economy is reflected in the materials of the cell wall and the physicochemical properties of the protoplastic elements and membranes which regulated their deposition.

The cell wall is a highly functional entity which varies in composition and architecture, internally in accordance with cell–cell and cell–tissue interactions, and, externally in accordance with environmental factors and stresses.

The student of ontogeny, whether concerned with cellular differentiation or morphogenesis in complex organisms, proceeds with confidence that a relationship indeed exists between form and function. The mechanisms of ontogenetic control constitute, however, one of the great frontier areas of quantitative biology. The nature of the interactions between the genomes of individual cells and biotic, physical, and chemical components of their environment are largely obscure, although the consequences of such interactions are always in evidence.

A study of the cell wall in its several aspects provides one approach to problems of cellular differentiation. In the pursuit of this study, consideration must be given to constitutional and architectural features of cell walls; to the material transformations associated with wall substance and the chemical and physical means for their regulation. Although the cell wall of the vascular plant has been selected as a major subject, a proper biological perspective requires comparative treatment of walls and kindred intercellular systems as they exist among organisms at large. Historical perspective is provided by examination of the phylogenetic aspects of a durable component of the plant wall, lignin. Among organisms, the walls of plant cells, particularly vascular forms, have been studied most extensively. Purely scholarly consideration of plant structure and growth have been reinforced greatly by long-standing economic and technological interests in cell wall derivatives. Accordingly some attention must be given to woods, plastics, fibers, and coal. Such considerations become all the more important when it is recognized that cell wall technology is a function of the fundamental chemistry, physics, and geology of these materials.

CELL TYPES AND CELL WALLS IN VASCULAR PLANTS

The identity of tissue types depends upon the anatomical characteristics of their component cells. Simple tissues contain but one cell type, complex tissues two or more.

Parenchyma, sclerenchyma, and *collenchyma* represent simple tissues, parenchyma is the primitive, unspecialized tissue consisting of isodiametric cells with active protoplasts and relatively thin walls. Parenchyma cells are the principle or sole constituent of meristems, pith, cortex, and other tissues. Collenchyma, which forms simple, homogeneous tissue, consists of irregular, elongate cells with unevenly but heavily-thickened walls. The walls are soft extensible and rich in cellulose and pectin. Collenchyma cells may be recognized by their wall pattern.

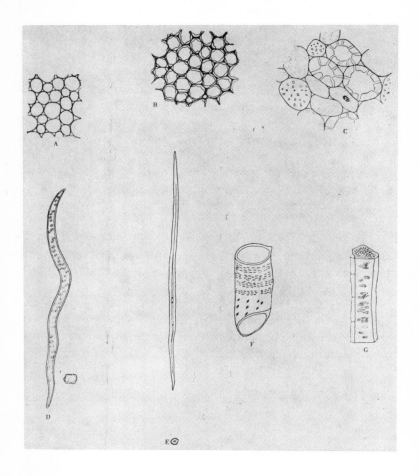

Fig. 1. Cell and wall variations in plant tissues. *A*, thick-walled Pith; *B*, collenchyma; *C*, sclerenchyma, showing sclereids and parenchyma cells; *D*, tracheid; *E*, libriform fibre; *F*, vessel element; *G*, sieve cell with companion cell. (From Eames, A. J. and Mac Daniels, L. H., *An Introduction to Plant Anatomy*, 2nd ed., pp. 83–96. McGraw-Hill, New York, 1947.)

Sclerenchyma, like collenchyma, is a supporting tissue. Unlike collenchyma cells, sclerenchyma cells have thick, hard, lignified walls and are low in water. During maturation their protoplasts degenerate, leaving a non-living sclerenchyma tissue. Sclereids and fibers are sometimes conveniently distinguished in sclerenchyma, but these forms are part of a morphological continuum. Fibers are found in many tissues, but most profusely in cortex, pericycle, xylem, and phloem.

The sculpture of fiber walls varies from the complex bordered pits of xylem to simple pits in the other complex tissues. Fiber cells may occur singly or in scattered clusters, but most commonly, they are present as strands or sheets of considerable length.

The anatomically defined fiber is not to be mistaken for seed hairs (cotton), foliar vascular bundles (hemp), wood cells (paper), etc., which are commonly termed fibers on the basis of their properties.

Sclereids tend to be isodiametric (but not exclusively so), although irregular and quite variable. They are hard, even gritty, and are sometimes known as stone cells. The pits in sclereid walls are simple and form branching cavity systems.

The anatomical array in the complex tissues—*xylem* and *phloem*—is far more varied than it is in the simple tissues. In xylem, *tracheids*, *fibers* (or fiber-tracheids), *vessels*, and *wood parenchyma* may be distinguished. Of these types, the tracheid occupies the most fundamental position. The tracheid wall is extensively pitted and, although the individual cells are closed off with tapered ends, their lumens are in communication through the delicate, sometimes perforated, pit membranes. This continuous, or anastomosing system can sometimes be demonstrated by the forced passage of carbon particles from cell to cell.

The primitive, rather "all-purpose" tracheid is present in some species, but it is often displaced by the apportionment of functions to the other cell types—support to fibers, conduction to vessels, and storage to parenchyma. Vessels possess both pits and end-wall perforations. The pits occur in those regions of the wall where vessels are contiguous. The end walls may contain simple perforations, scalariform (ladder-shaped), or, less commonly, reticular openings.

Like the pits, end-wall perforations involve highly selective lytic processes whose regulation and chemical nature is unknown. These sculpturings of the cell wall together with other wall modifications, give every appearance of being highly functional changes.

In phloem, as in xylem, parenchyma fiber and sclerid cells are present together with specialized conducting cells. The fundamental structural and functional unit is the *sieve element*. This cell is again distinguished by its wall sculpture—the presence of *sieve areas* and *sieve plates*. The former are clusters of fine pores through which cytoplasmic processes extend from one cell to its neighbors. The latter are perforated communicating regions generally confined to end walls. The pores are $0.5–3.5\mu$ in diameter. Just as vessel elements in xylem are joined end—to-end to form vessels centimeters or meters in length, the sieve tube elements are joined to form a continuous chain. Unlike the vessel elements, living cytoplasm is present in the cells although nuclei are lacking. A modified parenchyma cell, the companion cell, is associated specifically with the sieve cell elements. The nature of the relation between enucleate sieve elements and the nucleated companion cells is not known, but it is evident that they are closely associated in function as they are in position and origin from common mother cells.

These variations in cell type have been reviewed briefly to illustrate a part of the pattern of differentiation. Repeatedly, the condition of the cell wall characterizes these tissues and their component cell types which arise during differentiation.

The developmental plan which is encountered under mesic condition serves as a convenient reference point for the consideration and cell and wall specialization. When the physical environment deviates from these mild, "ordinary" conditions, however, the plant body must undergo suitable modification if it is to survive. In xerophytes, the place of the cell wall in adaptation is well illustrated.

Xeric environments, whether in desert, tundra, or saline regions, select for water-conserving modifications in their plant communities. Epidermal cells often exhibit the most conspicuous xeric adaptations although these modifications commonly extend into subepidermal layers, and even throughout the plant.

The development of thick cuticles, infiltration of cutin and waxes, and lignification of epidermal cell walls all represent devices for laying a water barrier over the plant surface. Cutinized or lignified hypodermal layers immediately beneath the epidermis are also found among the xerophytes. In some xerophytes, sclerenchyma with its hard, lignified walls is laid down to form insulating sheets or fiber strands.

Exceptional water resistance is found in cork. This tissue is the product of a specialized form of cambial activity. The cork cambium (phellogen) extends files of cells radially outward displacing cortical tissues. Differentiation in these cells entails the intercalation of lipoidal material (suberin) into the cell wall and deposition of lignin upon it. After protoplastic degeneration cork tissue consisting of water impermeable, air-filled cells remains.

The presence of increased amounts of superficial lignin and cutin not only provide water barriers, but may also serve as protective ultraviolet screens in regions of high light intensity.

Under hydric conditions, the problems of insulation, support, and radiation screening are minimized. Hence it is not surprising to find that the epidermis in typical aquatic plants possesses extremely thin cuticles permeable to water and gases, and that stems contain little or no lignin, even in vascular tissues, and no sclerenchyma.

The epidermal cell walls in seed coats sometimes resemble those described in xerophytes. The preservation of dormancy without desiccation for variable periods is often required in seeds. Accordingly, seed adaptations must include adequate, but temporary insulation for the embryo. In addition to cutinization and lignification, the seed coat exhibits a diversity of anatomical and chemical modifications which implicate the cell wall.

The few illustrations of wall and cell variations which have been presented show clearly that cell and cell wall form, function and constitution are indeed regulated by internal and environmental factors alike.

Keeping these concepts in mind, we may now examine the various aspects of cell wall science which must eventually provide a basis for understanding the regulation of cell wall formation by genetic and enviromental control factors.

CHAPTER 1

CONSTITUTION AND ARCHITECTURE IN THE CELL WALL

I. ANALYTIC PRINCIPLES AND PROCEDURES

THE study of cell walls requires the techniques of many classical disciplines and has stimulated the development of new ones as well. The traditional methods of organic chemistry together with newer biochemical methods yield considerable information about the polymeric components of the wall and their constituent subunits. In principle, however, the chemical approach is destructive, and can provide but little direct information about spatial relationships among these many substances. The development of physical methods which allow recognition of architectural elements has been of the utmost importance in the furtherance of cell wall studies. Physical methods are uniquely valuable as a means for probing the cell wall without disrupting it.

A full understanding of cell wall structure will depend upon the continuation of modern trends toward a combined physical–chemical approach.

We will now examine some of these analytic methods and approaches, and their operational basis. The methods and concepts which will be presented are designed to be illustrative rather than exhaustive. Following an inquiry into methodology, the products of these combined investigations will be employed in an effort to construct an integrated picture of the cell wall in its organized state.

Chemical Methods

Analysis of the cell wall by chemical means depends upon successful application of two kinds of technique. First, the several chemical classes represented must be separated from one another by methods which give a minimum of contamination or artifact formation; second, specific, and sensitive methods must be available for qualitative and quantitative measurements of the isolated fractions.

Dealing as it does with complex materials, natural products chemistry cannot achieve these technical ideals in a simple, direct fashion. It is only through the efforts, experiences, and ingenuity of many investigators that a definitive analytical description of the cell wall and similarly heterogeneous materials has been achieved. No brief account can hope to convey the difficulties with which this roadway is beset. Therefore, as we examine the current end-product of more than a century of cell wall chemistry, it is well to remember the efforts represented by the apparently "finished" character of the modern techniques and the likelihood of even further improvement in the future.

Among the simpler solvents, cold dilute NaOH may be used for the removal of high molecular weight carboxylic acids. Calcium salts of polymeric acids may be solubilized by treatment with ammonium oxalate or chelating agents which precipitate or sequester the calcium. More concentrated alkali solubilizes neutral polysaccharides at low temperature and waxy or other lipid materials when hot. A more specialized solvent for some acidic substances is warm dilute H_2O_2. Hot dilute acid also solubilizes some acidic components, but also hydrolyzes some neutral polysaccharides. Cellulose, the most resistant carbohydrate component, may be swelled and solubilized by cold sulfuric acid (72 per cent) or by cupraammonium solutions which contain the ion $Cu(NH_3)_4^{++}$.

The aromatic polymers of the cell wall may be partially extracted by organic solvents such as dioxane or hot butanol, or more completely solubilized by treatment with acidic alcohols. Strong alkali will also extract aromatic substances in proportion to their content of phenolic OH. More general (and destructive) extractants for aromatic compounds include bisulfites and chlorine in combination with H_2SO_3 or ethanolamine.

By the proper application of solvents and treatments, the cell wall may be fractionated into cellulose and the following additional major chemical classes:

(1) Non-cellulosic polysaccharides.
(2) Pectic substances.
(3) Polyuronide hemicelluloses.
(4) Lignin.

In addition, cuticular substances, mucilages, tannins, pigments, and terpenoid compounds are present in various amounts.

It is not possible to obtain clear separation and maximum recovery of each of the major components by means of a single fractionation scheme. Hence, some isolation procedures must be designed for quantitative recovery of one or a few components in essentially pure form at the sacrifice of other constituents, which must, then, be approached by other techniques.

Fractionation procedures may sometimes be facilitated by enzyme treatments. Thus, the pectic enzymes may be used for removal of the various pectic compounds, and cellulase can be applied for dissolution of cellulose. Such procedures are most useful as pretreatments to render other components, especially lignin, more accessible to milder solvent extractions. More generalized microbial attack upon cell wall materials may lead to removal of all but the most resistant substances such as lignins and cuticular materials.

After the cell wall components have been separated and recovered in purified form, they may be identified and analyzed.

Cellulose is commonly identified as a residual substance after removal of other wall components by successive chlorine–Na_2SO_3 and alkali treatment. This virtually pure residue may be further analyzed for contaminants and ash, or may be dissolved in cuprammonium, reprecipitated by acid, and weighed as purified cellulose.

The non-cellulosic polysaccharides include a variety of carbohydrate polymers whose specific analytical properties differ. Collectively, they may be determined, however, after treatment of the whole cell wall with chlorine combined with Na_2SO_3 or ethanolamine as the fraction soluble in cold 17.5 per cent NaOH.

The pectic substances are determined after removal of polyuronide hemicelluloses with cold 4 per cent NaOH, by solubilization in 0.05 N HCl, alkaline hydrolysis, and alcohol precipitation as calcium pectate. Residual pectic substances which sometimes remain are solubilized with hot ammonium oxalate and after separation from the oxalate by precipitation, may be redissolved in ammonia and reprecipitated as the calcium salt.

As we have noted, the hemicelluloses are distinguished by their solubility in cold dilute alkali, a property which provides the basis for their determination. The hemicelluloses may be recovered from alkaline extracts by precipitation with cold acid and alcohol. Commonly, alkali extraction is preceeded by treatment of the tissues with ammonium oxalate in order to remove the pectic substances.

Fractional precipitation has been used in the separation of various polysaccharides, the hemicelluloses especially. Fractionations have been carried out with acids, alcohol, or by complexing with copper. Cetyltrimethyl ammonium bromide will precipitate acidic polysaccharides as insoluble quaternary ammonium salts, leaving neutral polymers in solution. Even the latter may be precipitated if first rendered acidic by formation of borate complexes.

Among the major cell wall substances lignin is distinctive for its aromatic character and the relative difficulty encountered in separating it without alteration from other components. The many separation techniques which have been developed range from the delignification of wood for practical ends (in sulfite pulping processes, for example) to mild treatment of the cell wall with cellulose to render the residual lignin more soluble.

Traditional procedures are based principally upon the removal of other substances leaving the lignin as a residue. The resistance of lignins to strong mineral acids permits separation of carbohydrate derivatives which are solubilized by 72 per cent sulfuric acid or fuming (40–42 per cent) HCl. Although such procedures modify the lignins they are probably suitable for gross analyses. A milder isolation procedure is based upon treatment of cell walls with dilute alkali, hot dilute acid, and cuprammonium solution. There are some losses of lignin in alkaline solutions, but a more native product may be obtained.

The reverse procedure, removal of lignin from other wall constituents, has been accomplished using bases such as alcoholic NaOH or ethanolamine; phenols; or alcoholic HCl. Again, such reaction solvents do not yield a native product, but they are useful in the study of lignin structure.

After comparatively pure substances have been isolated from the cell wall, their constitution may be examined. The polymeric nature of the major wall components has long been recognized even if their study has only recently been undertaken from the viewpoint of polymer chemistry. In polysaccharide chemistry, one of the foremost analytical procedures is hydrolysis. Acid hydro-lysis degrades polysaccharides, yielding mixtures of lower poly-saccharides, oligosaccharides, and simpler units. Ultimately, the most stable structural units, the monosaccharides may be obtained, but it is also of value to carry out controlled hydrolyses in the hope that relatively stable intermediates, disaccharides for example, may be identified.

Constituent monosaccharides may of course be identified by means of a variety of chemical and physical properties including optical activity, melting points and crystallographic features of derivatives (osazones formed with phenylhydrazine, for example), titrable acid groups when present, etc. The application of various analytical procedures may be complicated when two or more constituent units are present in the hydrolysate. Specific tests may sometimes suffice to demonstrate the presence of more than one carbohydrate but recourse to separation procedures is also necessary. To the more classical methods of separation based upon physical differences (solubility, for example) among sugars or their derivatives, the techniques of paper and column chromato-graphy may be added.

The polyfunctional nature of the sugars allows the existence of a variety of combinations in polysaccharides. The most important variables are made manifest by consideration of a single mono-saccharide unit in the polymer. Thus, we must determine: (a) The number of units linked to a specific residue (linear vs. branched); (b) the frequency of end residues (which contain one more hydroxyl than an internal residue); (c) and the mode of linkage to adjacent residues (which includes both the distinction between α- and β-isomers and the alcoholic carbon atoms through which linkages are established).

Enzymes have found application in the determination of α- and β-linkages. The α-glucosidase, maltase, and the β-glucosidase, emulsin, may be cited as analytically useful enzymes specific for their respective isomers. The most useful analytical technique which has been applied to these relational aspects of polysaccharide structure is methylation prior to hydrolysis. Commonly, dimethyl sulfate in alkali has found use in this connection. A linear glucose homopolymer containing 1,4-linkages will yield principally 2,3, 6-trimethyl glucose. Short polymer chains will yield appreciable amounts of 2,3,4,6-tetramethyl glucose which is derived from frequent end groups, whereas only trace amounts of the tetramethyl derivatives can be isolated if the polymer is large and possesses a small proportion of end residues. The presence of branching is shown by the occurrence of 2,3-dimethyl glucose. Additional evidence about linkages may be obtained by methylation of acid derivatives or bromine oxidation.

Sugars other than glucose released by hydrolysis are often identified by specific reactions. Mannose may be estimated by precipitation of its phenylhydrazone. Galactose is readily identified by oxidation (nitric acid) to water insoluble mucic acid. In general, pentoses (xylose, arabinose) are identified and estimated by dehydration (12 per cent HCl) to furfural and formation of insoluble colored products with reagents such as phloroglucinol.

The degradation of lignin, like its isolation, presents distinctive problems. Among the many analytical procedures, alkali fusion, hydrogenolysis, various forms of pyrolysis, and oxidation with permanganate or nitrobenzene are of considerable importance. These chemical methods coupled with ultraviolet and infrared spectrophotometry have contributed considerable knowledge about lignin structure. Methylation procedures have also been applied so degradation products for the identification of phenolic groups, and hydroiodic acid treatment is useful in the determination of methoxyl groups.

Most histological staining methods which have been applied in studies of various plant cells and tissues lack the specificity of reliable microchemical reagents. For example, standard dyes such as hematoxylin, and safranine, have been claimed as "cellulose" and "lignified wall" reagents, respectively. The great dependency of staining reactions upon conditions and the frequent reversal of staining affinities cast doubts upon the general value of these dyes. Such staining reactions have undoubted utility in cytological investigations, but may be better indicators of the organization and polyelectrolyte character of macromolecules and their aggregates, than of specific chemical groups. This view is supported by dyes such as azure B, which exhibit the phenomenon of metachromasy, or differential coloration in different structures. One dye, ruthenium red, seems to be of comparatively great utility as a differential stain for pectic substances. Although it cannot be claimed that this dye is specific for pectins, it seems to have particular affinity for highly polymerized acidic carbohydrate derivatives of this type.

Specific cellulose staining can be effected by virtue of its molecular organization. The two most useful reagents are chlorozinc iodide and KI_3–72 per cent sulfuric acid. These reagents swell cellulose and exhibit dichroism, appearing blue or violet when the electric vector of incident light is parallel to the axis of the fiber and colorless when the electric vector is normal to the fiber axis.

Such effects are based upon the oriented deposition of iodine within the ordered fiber structure. The staining of cuticular materials also depends upon a physical property—solubility—in lipophilic dyes such as sudan III.

Lignin is unique among major wall substances in its ability to form chromogenic condensation products when treated with acidic solutions of many phenols and aromatic amines. When suitable precautions are taken to remove soluble, low molecular weight interfering substances, reagents such as phloroglucinol–HCl can be used to impart a brilliant red stain to lignins and lignified walls. Other lignin-specific chromogens may be formed by treatment of tissues with chlorine followed by alkali or sodium sulfite.

Physical Methods

High polymers such as cellulose are organized into reticular gel systems. The reticulum consists of variously ordered molecular aggregates, micelles, together with intermicellar substances. Physical methods which have been employed in cell wall studies are directed toward demonstration of the fine structure in reticular systems and the texture, or orientation of structural components.

In principle, the successful application of optical methods to the study of micellar frameworks depends upon the relation between micellar strands and intermicellar space and upon the presence of some degree of spatial orientation.

Ideally textured bodies consisting of parallel isotropic cylinders or parallel isotropic planes (composite bodies), become optically anisotropic if the diametres of the rods and the distances between rods or layers are small compared to the wavelength of light, and if there is a true phase boundary between the structural units and the material in the spaces separating them. A homogeneous crystalline substance which exhibits the same index of refraction from all aspects (that is, crystallographic axes) is isotropic, whereas such a substance which refracts light differently along different axes is double-refracting, or anisotropic. If perfectly isotropic rods or planes are oriented in a suitable dispersing medium, the system of isotropic rods or planes as a whole will rotate plane polarized light which is passed through it if the medium enclosed by the particles differs from the particles in its refractive index. Further, the magnitude of double refraction is a hyperbolic function of the refractive index of the dispersing medium.

As this kind of anisotropic behavior decreases to zero when particle and dispersing medium have the same refractive index, it depends upon the oriented condition of the particles, that is, upon the texture of the composite body, hence may be termed "textural double-refraction". Textural double refraction has sign as well as magnitude. Thus in a fiber or thread which consists of rodlets oriented parallel to the long axis of the fiber (or "fiber axis") the double refraction is positive, whereas a fiber made of stacked plates whose planes lie perpendicular to the fiber axis has negative double refraction.

As we have noted, homogeneous crystalline materials may show double refraction. Similarly, the long chain molecules constituting the strands of a gel will exhibit double refraction which is not abolished by changing the refractive index of the dispersing liquid. In such substances optical anisotropy is an intrinsic molecular property, hence it is called intrinsic double refraction and may be expressed as the difference between refractive indices parallel and perpendicular to the fiber axis. The intrinsic property also has sign as well as magnitude. In all, therefore, six types of double refraction which arise from combinations of textural and intrinsic double refraction may be distinguished.

The anisotropic properties of gels are further complicated by their elastic properties. Elastic deformation of gels under tension gives rise to tension double refraction which is positive with respect to the axis of deformation and compression double refraction which is usually negative. Beyond the elastic limits, these characteristically reversible optical features give way to new, more intensely anisotropic properties as a result of orientation under plastic deformation.

When crystalline materials are placed between a monochromatic X-ray source and a photographic plate, the X-rays are diffracted and produce characteristic patterns on the plate. Polymeric substances which form crystalline aggregates with large numbers of lattice planes show characteristic interference patterns. The lines and spots recorded can be analyzed for distance, density, width, and arrangement of interferences. From such measurements we can, in turn calculate: (1) distances between lattice planes, particularly the fiber period; (2) the number of atoms in the planes; (3) the width of undisturbed lattice planes, and (4) the arrangement of rod-shaped lattice regions. X-ray diffraction can only be applied to substances or regions containing an ordered lattice. Amorphous intermicellar substances do not give X-ray interferences, hence their study by X-ray diffraction can only be effected by introduction of a crystalline material into these spaces. For this purpose, gold and silver salts are introduced and reduced by light or chemical means. From X-ray interference rings obtained with gels so treated, the size of the cubic gold and silver crystals, hence of the intermicellar spaces, may be calculated.

Among useful optical methods of analysis other than those considered, electron microscopy ranks high. High-voltage electron beams possess very short wavelengths, of the order of 0.005 mμ, when compared with light and even hard (high-voltage) X-rays. Accordingly, the resolving power of the electron microscope is far greater than that of the light microscope. Electron microscopy has provided eloquent proof of the reticular structure of various gels as inferred from indirect measurements.

A quite different physical characterization of complex substances is represented by measurement of their hydrodynamic properties. When solutions of polymers are forced through fine orifices (capillary viscometers) or subjected to a torque (cylinder viscometer) their response to shearing forces reflects their size shape and molecular interactions. In the ideal hydrophilic sol, viscosity results from the interaction of single molecules with solvent, hence a molecule consisting of repeating units will interact with solvent in relation to its degree of polymerization. Very great size (length) or insufficiently diluted solutions lead to interactions among solute molecules as well which complicate interpretation of viscous behavior. Such interactions lead to particles which no longer form a true sol consisting of single molecules, and, finally to gelation. The molecules in colloidal solutions can also be characterized by their behavior in a gravitational or centrifugal force field.

The displacement of particles (sedimentation) subjected to enormous angular accelerations ($10^5 \times g$) in the ultracentrifuge is determined by their effective particle weight. Ideally, the particles of the sol consist of single molecules, hence, with suitable correction for densities of particle and medium, sedimentation may be related simply to molecular weight.

When colloidal particles also carry a charge, they may be displaced in solution by an applied electrical field. Electrokinetic behavior, or electrophoresis, is characteristic of acidic wall polymers such as pectic acids or agar, but not of cellulose or other neutral polysaccharides. Electrophoresis depends upon hydrodynamic and properties, hence the proper analysis of displacement in an electrical field must take account of viscous behavior as well.

II. THE RESULTS OF ANALYSIS

AN accurate accounting of cell wall components depends for its construction upon the thoughtful application of the chemical and physical methods of analysis which have been discussed. Although there are, undoubtedly, many improvements yet to be made in the particulars of cell wall analysis, the greatest need lies in the interpretation of data and the development of an integral picture of the cell wall as an organized system.

Before attempting to make a synthesis of the various kinds of analytical data at our disposal, we must of course, examine more closely the information which these data yield about the constitution and organization of the polymers and other molecules which provide the elements of cell wall structure.

Fractionation of the Whole Cell Wall

Before examining a separation scheme, we may note by way of review, the salient solubility properties of individual wall constituents (Table 1). Cellulose and the cuticular substances are notably more resistant to an array of solvents than are the other components, lignin included. Lignin, the cuticular materials and cellulose are notably acid resistant.

Although a simple, universal fractionation procedure is obviously desirable, the intergradation of properties, and the great variability which is to be found in the wall components of different species or even in walls which are in different stages of differentiation renders such an aim impossible or nearly so. Further, interest in the isolation or removal of particular wall components in specific tissues or organs has stimulated the development of many procedures which are too gross for accurate, complete analysis. The Cross and Bevan treatment is an example of an historically important procedure. Whole cell wall material (wood for example) is treated with chlorine and sulfites with the resulting solubilization of lignin and polyuronides. The residue, known as Cross and Bevan cellulose actually consists of true cellulose together with non-cellulosic polysaccharides. Pure α-cellulose can be obtained from this residue by extraction of the non-cellulosic component with cold 17.5 per cent NaOH.

TABLE 1. SOLUBILITY CHARACTERISTICS OF WALL COMPONENTS

Reagent	Cellulose	Non-cellulosic polysaccharides	Polyuronide hemicelluloses	Pectic subst.	Lignin	Cuticle subst.
Cold dil. alkali	–	±	+	–	–	–
Hot. dil. alkali	–	+	+	+	±	±
Hot. conc. alkali	+	+	+	+	+	+
Hot. dil. acid	–	Hydrol	Hydrol	+	–	–
Cold 72% H_2SO_4	+	Hydrol	Hydrol	Hydrol	–	–
Hot dil. alcoholic KOH	–	X	Partial	X	Partial	+
Cold conc. alkali	–	+	+	+	–	±
Warm dil. H_2O_2	–	–	–	+	Oxid.	–
Warm dil. ammon. oxalate	–	–	–	+	–	–
Cuprammonium solution	+	+	–	–	±	–
Ethanolamine	–	–	–	–	+	–
Dioxane, phenols	–	–	–	–	+	Partial
Hot. dil. alcoholic HCl	–	X	X	X	+	–

A more generally useful scheme begins with the removal of lignin by extraction of whole cell wall with chlorine and hot ethanolic ethanolamine solution. The residue is then treated with cold 4 per cent NaOH to remove the polyuronide hemicellulose, and finally cold 17.5 per cent NaOH is used to separate cellulose (residue) and the non-cellulosic polysaccharides. Unfortunately, some loss of non-cellulosic polysaccharide also occurs in dilute alkali, necessitating an additional independent determination of this fraction.

Methods such as the foregoing which have developed out of wood technology do not take account of the small quantities of pectic substances present in woody tissues. The pectic substances may be separated independently by removal of interfering polyuronides with cold 4 per cent NaOH followed by the procedure already described.

Upon comparison of cell walls from the organs of various plants, one is immediately impressed by the great differences in the proportions of their constituents (Table 2). Relative to the other major substances, cellulose shows somewhat less variation among the samples given. In contrast, the non-cellulosic polysaccharides, pectic substances, and polyuronide hemicelluloses vary by as much as seven- to sixty-fold.

TABLE 2.   PERCENTAGE CELL WALL COMPOSITION IN VARIOUS PLANT ORGANS

Component	Maize coleoptile	Sunflower hypocotyl	Maize straw	Maize cob	Hop flower	Bamboo stem	Pine sapwod
Cellulose	36	38	46.5	38.3	30.8	41	53.6
Non-cellulosic polysaccharide	30	8	8.3	0.5	5.4	10	8.8
Pectic substances	13	46	0.3	0.5	—	—	1.0
Polyuronide hemi-cellulose	—	—	33.4	42.4	9.7	14	3.0
Lignin	—	8	19.5	16.7	54.1	32	26.4
Cuticular substance	21	—	—	—	—	—	—
Protein	—	—	1.9	3.2	—	—	—
Ash	—	—	3.5	1.4	—	—	—

In spite of these extreme variations in individual non-cellulosic components, their sum has no more range than does cellulose itself. This relationship suggests that in the cell wall the relatively constant cellulose is associated with a relatively constant interstitial aggregate whose composition varies, perhaps in a functional and dynamic fashion. We shall have occasion to return to this proposition later. Analytical data on cell wall substance is of vital importance to an understanding of biosynthetic processes, particularly in cellular growth and differentiation. The serious consideration of detailed analyses is, however, hampered by the coexistence of data obtained over a span of years by the use of fractionation procedures which have themselves been subject to revision and improvement. Rarely has there been opportunity for a critical comparison of older and newer methods for cell wall fractionation. A particularly important comparison does exist, however, for the Avena coleoptile, an organ of critical importance to those who study the physiological chemistry of growth (Table 3). In analyses separated by a quarter century, major differences in all principle wall components exist. The most significant distinction lies in the pectic substances which have been specifically implicated in the hormonal regulation of cell extension. In the older analysis, the pectic substances form an appreciable part of the wall, whereas the more recent analysis demonstrates only trace amounts. It has been claimed that ammonium oxalate solubility, which is commonly

TABLE 3.   CONSTITUTION OF THE COLEOPTILAR CELL WALL OF AVENA

Component	Thimann and Bonner (1933)	Bishop *et al.* (1958)
Cellulose	42	24.8
Non-cellulosic polysaccharide	38	50.7
Pectic substances	8	0.3
Polyuronide hemicelluloses	—	—
Lignin	—	—
Waxes, fats, pigments	—	4.2
Proteins	12	9.5

adduced as evidence for the presence of pectins may be misleading, as polysaccharides other than pectin may also appear in this extractant in appreciable quantities. The significance of this major discrepancy for the interpretation of growth phenomena will be examined further in the discussion of cell wall dynamics, but the warning that it carries should not pass unnoticed.

Wall Polysaccharides

On complete acid hydrolysis, cellulose is converted quantitatively into β-D-glucose. Controlled hydrolysis yields the disaccharide cellobiose, 4-D-glucopyranose-β-D-glucopyranoside. Hydrolysis of methylated cellulose yields 2,3,6-trimethylglucose, and no dimethylglucose. Careful anaerobic methylation yields no detectable tetramethylated product. From these facts, cellulose may be pictured as consisting of extremely long, unbranched chains of anhydroglucose residues joined through 1,4-linkages.

The crystallinity of cellulose has been clearly established by X-ray diffraction planes normal to the axis at 10.3 Å intervals, and parallel to the axis at 8.3 and 7.9 Å intervals. The crystallographic unit cell is rhombohedral, measures $10.3 \times 8.3 \times 7.9$ Å; and has a characteristic angle of 84° between the shorter sides. From its dimensions, we may calculate that the unit cell has a volume of 675 Å^3. This volume multiplied by the density of cellulose (taken at 1.6 g or 1.6×10^{-24} g/Å^3), gives a unit cell mass of 1.08×10^{-21}. This mass divided by the weight of the anhydroglucose residue ($C_6H_{10}O_5$) of about 162 gives moles of residue/unit cell, 6.7×10^{-24}. The latter multiplied by Avogadro's number 6.06×10^{23}, yields a number of glucose residues very close to 4. The 10.3 Å identity period corresponds to the distance between the number one carbon of one residue and the oxygen at the number four carbon of the second residue in cellobiose. The 7.9 Å spacing agrees satisfactorily with the maximum width of the cellobiose residue, 7.5 Å. Thus, the unit cell contains cellobiose units lying in planes 8.3 Å apart. The unit cell is best depicted with four cellobiose residues lying on the four edges of a parallelepiped, each such dimer being shared by the four unit cells to which that edge is common. From the number of glucose residues per unit cell, an additional cellobiose chain must extend through the center of the parallelepiped, but is out of phase by one-half residue with the other and antiparallel to them.

We have now examined the main features of crystalline structure, but the unit so defined gives no indication of the length of the individual polymer chains. In the absence of oxygen, cellulose may be dissolved in cuprammonium solution, that is, dispersed into individual molecular chains, without appreciable degradation.

The molecular weights of these chains have been determined by viscometry or sedimentation measurements in the ultracentrifuge. From such determinations, the molecular weight of cellulose is found to vary with the source, and methods of preparation and measurement, but is consistently high:

Reported variations notwithstanding, cellulose has a degree of polymerization of 1400–10,000, hence molecules ranging in length from 7000–50,000 Å, (0.7–5 mμ).

The ordered structure of cellulose is also revealed in its optical properties. Isolated cellulose exhibits intrinsic double refraction and dichroism. By taking the utmost precautions for the exclusion of oxygen during cuprammonium extraction, a molecular weight for cellulose of over 20,000,000 has been obtained, corresponding to a degree of polymerization of 100,000 or more, and a molecular length of at least 0.05 mm.

Source	Method	Molecular weight	Degree of polymerization
Cotton	Visc.	330,000	2020
	Ultracent.	150,000–500,000	1000–3000
	Ultracent.	1,500,000	9200
Ramie	Visc.	430,000	2660
	Ultracent.	1,840,000	11,300
Hemp	Visc.	320,000	1990
Spruce	Visc.	220,000	1360

The molecular weights of pectic and pectinic acids vary considerably with source, measurement technique and isolation method. There is some question still about the intactness of isolated pectins, hence the reported molecular weights may represent a greater or lesser fraction of the native molecule. They may, however, be taken as minimum values. Among the reported molecular weights are: apple pectin, 67,000; apple pectic and pectinic acids, 62,000–280,000; lemon pectin, 89,000; beet pectin 62,000; beet pectinic acid, 90,000; and flax pectin 64,000.

The pectins are characterized by their content of ester methyl groups. The ideal limits for these polymers are unesterified polygalacturonic acid and the fully esterified molecule which has no acidic properties. The intermediates with varying proportions of free and esterified carboxyl groups represent the pectic and pectinic acids. The maximum methoxyl content for pectins containing twenty-five or more residues is 16.3 per cent. The actual methoxyl contents vary from the low ester range, 3–7 per cent, to high values of 12–15 per cent. As the methoxyl content is lowered, the pectic substances become more sensitive to di- or poly-valent cations.

The constitutional differences between protopectin and the other pectic substances remain unclear. Empirically, the protopectins are distinguished by their insolubility, and, in general, by a higher molecular weight. To account for their unique properties it has been proposed that protopectin is associated with cellulose or polyvalent cations, or has an intrinsically greater molecular size. One aspect of ion–pectin interaction of some interest involves the insolubility of double salts with D-galacturonic acid; solubilities are ranked: $Na^+ \gg Ca^{++} > Ca^{++}$, Na^+. Double salts which form the most insoluble combinations could easily be provided in biological systems. In essence, pectic acids are poly- (anhydro-α-D-galacturonic) acids formed as is cellulose by 1,4-linkages. The presence of other linkages is still a matter of uncertainty. From X-ray diffraction studies, it has been established that the pectic acids differ in spatial arrangement from other polysaccharides such as cellulose or alginic acid. The examination of fibers prepared from sodium pectate and pectic acid shows that a right angle is formed by the glycosidic C–O bond and the plane of the galactopyranose ring, whereas the comparable angle in cellulose is only 20°. The ring planes must, therefore, be *trans* to one another, and the galacturonide chain must contain a three-fold screw axis. The identity (fiber) period is 13.1 Å. This structure suggests a molecule with more rigidity than is found in cellulose or alginic acid.

In general the molecules are 730–5400 Å in length, but regularly 10 Å in thickness. Dimensional variations exist, according to the biological origin and physical method employed. Thus, pectinic acid (the partially esterified acid) from apple has dimensions of 1200 Å × 10 Å based upon sedimentation velocity, but 2630 Å × 7 Å when measured by streaming birefringence (double refraction). By viscometry, an intermediate length of 1700 Å is found. The macromolecule in lemon peel measures 1600 Å × 10 Å in the ultracentrifuge, but 3780 Å × 7 Å by streaming birefringence. The removal of pectinic acid methyl ester groups to yield pectic acid obviously alters the overall size of the molecule, principally by allowing increased hydration, but leaves the general picture of a comparatively elongated form unchanged.

The length–diameter ratio given by the foregoing measurements are 120–377 (apple) and 160–540 (lemon). From intrinsic viscosity data, values of 53–165 have been obtained.

The celluloses and pectic substances exist, as linear polymers which are variable in molecular size, but uniform with respect to monomer type and linkage. The properties of the pectic substances vary widely according to the degree of methylation and of acid groups and their interaction with polyvalent cations. Such differences are minor, however, relative to their uniform identity as polygalacturonic acids. In turning, now, to the polyuronide hemicelluloses, we are confronted with polydispersity, branching, and monomer heterogeneity all at once.

The acidic hemicellulose fraction which is isolated in 4 per cent NaOH is not homogeneous physically or chemically. When the extract is acidified (acetic acid), the hemicellulosic fraction A precipitates. This fraction consists largely of linear polymers with few end groups. When the remaining solution is mixed with ethanol or acetone, fraction B, which contains branched polymers, precipitates. The still soluble component is fraction C. Each of these fractions may in fact be multicomponent as shown by subfractionation by stepwise addition of acid and alcohol. Other successful fractional precipitation methods based upon complexes or organic salts have been noted.

Hydrolysis of the acid hemicellulose yields a mixture of pentose and uronic acids. Of particular note are the aldobiuronic acids, disaccharides whose linkage is resistant to hydrolysis. The fundamental monomers are xylose, arabinose, glucuronic acid, O-methylglucuronic acid galacturonic acid, and occasionally rhamnose.

The stability of the aldobiuronic acids points to their importance as recurrent structural elements of the polymers. The constitutional relationships among these acidic disaccharides vary considerably with the source (Table 4). This tabulation includes for contrast, some hydrolysis products more correctly associated with mucilages and gums.

The uronic acids always occur as end groups or single unit side chains. As side groups, they are commonly attached to a D-xylose backbone chain at position C–2, sometimes at C–3.

A more complete picture of specific hemicelluloses can be given by examination of some which have been isolated. Thus, a homogeneous maize seed coat hemicellulose soluble in lime water gives on hydrolysis:

TABLE 4. HEMICELLULOSIC ALDOBIURONIC ACIDS

Uronic Acid	Aldose	Linkage	Source
α-Glucuronic	D-αxylobiose	1,4	Maize (cob)
α-Glucuronic	D-Xylose	1,3	Wheat (straw), pear (fruit), sunflower (head)
4-O-Methyl-α-glucuronic	D-Xylose	1,2	Beech, birch, elm, pine, cedar, hemlock, maize, wheat
4-O-Methyl-α-glucuronic	D-Xylose	1,3	Pine, wheat, (straw)
4-O-Methyl-α-glucuronic	D-αxylobiose		Cottonwood
3-O-Methyl-α-glucuronic	D-Xylose	1,3	Jute
β-Glucuronic	D-Mannose	1,2	Gums (cherry, damson)
β-Glucuronic	D-Galactose	1,6	Gums (acacia, almond)
Galacturonic	L-Rhamnose	1,2	Mucilage (elm, flax, plantain)
4-O-Methylglucuronic	L-Arabinose	1,4	Gum (lemon)

L-arabinose

D-xylose
D-galactose
L-galactose
β-D-galactopyranosyl-1,4-D-xylose

D-α-xylopyranosyl-1,4-D-α-glu-
copyran-osyluronic acid
D-xylopyranosyl-1,3-L-arabinose
L-galactopyranosyl-1,4-D-xylo-
pyranosyl-1,2-L-arabinose

The molecule containing these units has a xylan nucleus with branch elements consisting of xylo-arabinose, galacto-xylo-arabinose, glucuronic acid, and D-galactose. Both D- and L-galactose also occur as end groups.

One of the several maize cob hemicelluloses consists of highly branched molecules, degree of polymerization (*DP*) 150, containing about 59 per cent D-xylose, 22 per cent L-arbinose, 11 per cent methylglucuronic acid, and 8 per cent D-galactose. Glucuronic acid and galactose occur only as non-reducing end units. There are ten branch chains for every forty-four-unit linear chain, and one glucuronic acid end group for each nine to eleven sugar residues.

Diagrammatically, a wheat straw hemicellulose, DP 71–76, is constructed in the following manner:

```
xylose 1—4 xylose 1—4 xylose 1—4 xylose 1—4 xylose
                3          3          3
                |          |          |
                1          1          1
xylose 1—4 xylose     glucuronic   arabinose
                          acid
```

Among the hemicelluloses of beech species, two quite distinct polymers which have been described are:

```
xylose 1—4  xylose 1—4  xylose 1—4  xylose 1—4  xylose
                                              2
                                              |
                                              1
                              methylglucuronic acid
```

and

```
xylose 1—4  xylose 1—4  xylose 1—4  xylose 1—4  xylose
          2                                 2
          |                                 |
          1                                 1
      xylose 1—4 xylose        methylglucuronic acid
```

By comparison with celluloses with molecular weights ranging up to 10^6 or more and pectic substances in the range of 60,000–300,000, the acid hemicelluloses are relatively small molecules. The degrees of polymerization and approximate molecular weights ($DP \times 160$) of some representative polyuronides are:

Hemicellulose	DP	Mol. wt.
Oat straw	40–45	7000
Wheat straw	71–76	12,000
Flax straw	135	22,000
Maize cob	150	24,000
Beechwood	50–77	8000–12,000
Birchwood	121–209	19,000–33,000
Sprucewood	96–102	15,000

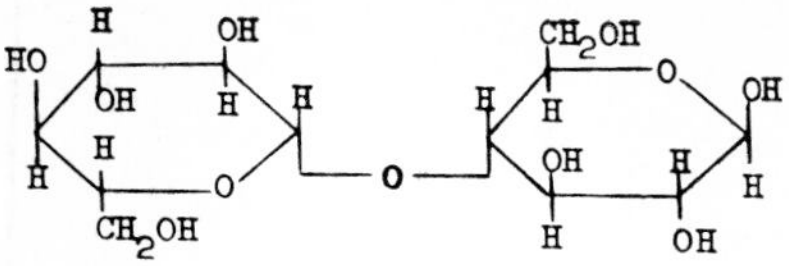

PLATE I

Cellobiose

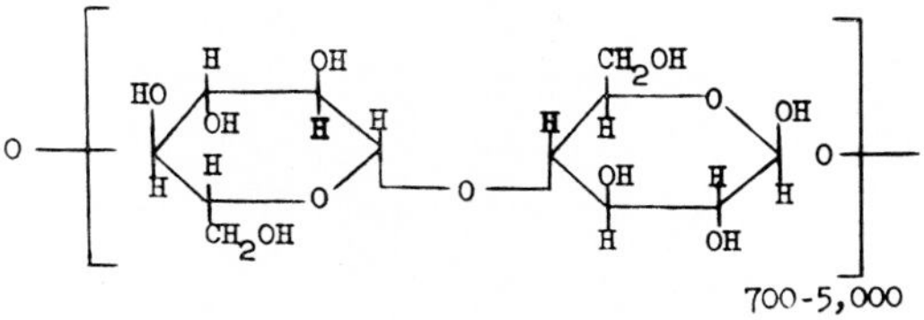

Cellobiosic unit of cellulose

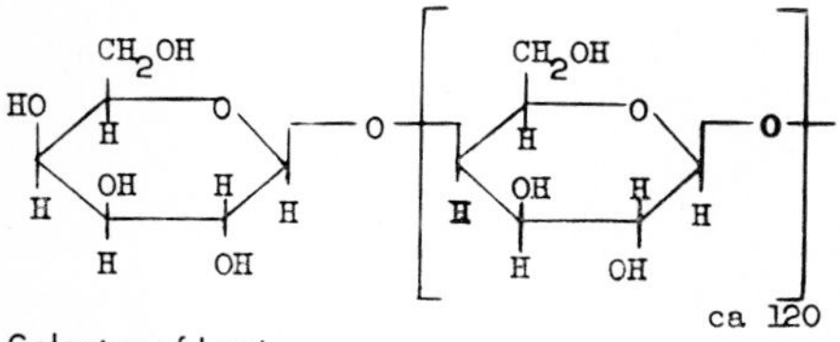

700-5,000

Galactan of Lupin

ca 120

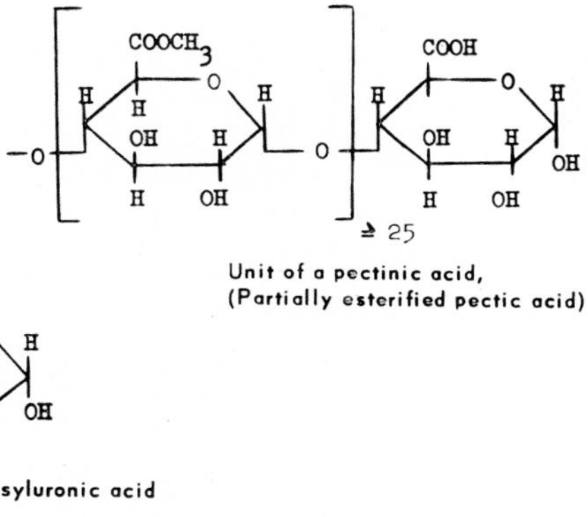

Unit of a pectinic acid,
(Partially esterified pectic acid)

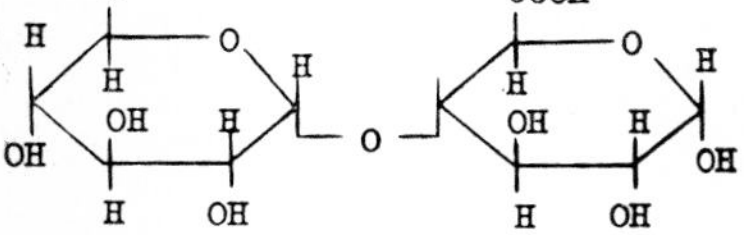

≥ 25

D-α-xylopyranosyl-1, 4-D-α-glucopyranosyluronic acid

Examples of the units of cell wall Polysaccharides

There is a natural gradation between the polyuronide hemicelluloses and the non-cellulosic polysaccharides. Indeed, both groups are sometimes treated collectively as acidic and neutral forms of hemicellulose. The major hemicellulose of the maize cob, which constitutes three-fourths of the alkali-soluble components consists of a linear xylan with only one glucuronic acid per molecule. Such polymers will not differ greatly from xylans themselves. Among these neutral hemicelluloses or non-cellulosic polysaccharides, there are homopolymeric mannans, galactans, arabans, xylans, and copolymeric glucomannans, galactomannans, arabinogalactans and arabinoxylans. These polymers which may be termed "glycans" vary considerably from species to species. Representative glycan contents of various plants are:

		Mannan (% whole)	Galactan (% whole)	Xylan (% of Cross and Beaven cellulose)
Woods:	spruce	6	—	—
	juniper	2.4	—	—
	sequoia	2.6	0.4–0.5	12
	pine	—	0.3–1	—
	oak, beech, ash	—	—	20–25
Fibers:	hemp	—	—	19
	ramie	—	—	2
Straws:	oat, barley, wheat	—	—	25–29

As in the acidic polymers the neutral compounds show diversity in residues, linkages, branching and size. Ivory nut mannan is a linear homopolymer of $DP=75$ in which the residues are joined by 1,4-linkages. A high degree of crystallinity is indicated by X-ray diffraction. Lupin galactan is also a cellulose analog, consting of linear 1,4-linked β-D-galactose chains.

Hardwood arabinogalactan is a branched polymer of molecular weight 16,000 for which the following structure has been proposed:

```
—6 galactose 1—6 galactose 1—
        3                    3
        |                    |
        1                    1
galactose 6—1 galactose 6—1 galactose 6—
        3                    3
        |                    |
        1                    1
    galactose 6—1 arabinose
        3
        |
        1
    galactose 6—1 galactose
```

This polymer may be accompanied by a branched galactan, molecular weight 100,000 containing β-1,3 linkages.

A branched corn cob polymer of molecular weight 13,700 contains 10.6 per cent L-arabinose with 89.4 per cent D-xylose and has been formulated as:

```
xylose 1—4 xylose—4 xylose—4 xylose 1—
         3         3         3
         |         |         |
         1         1         1
    arabinose arabinose  arabinose
              2          2
              |          |
              1          1
          arabinose     xylose
```

The glucomannans of conifers appear to be β-1,4-linked linear polymers ranging from 70 to 140 in degree of polymerization.

Still a different linkage system may be found in the arabans commonly isolated from pectin preparations. Both apple and peanut yield a branched arabinose homopolymer which may be formulated as:

```
—5 arabinose 1—5 arabinose 1—
                3
                |
                1
            arabinose
```

The gums and mucilages include occasional cell wall polymers, and many close relatives. They follow the pattern exhibited by other non-cellulosic compounds, having both neutral and acidic types. Acidic components include L-glucuronic acid, D-galacturonic acid, D-glucuronic acid, sulfate and phosphate groups. The neutral sugars associated with neutral and acidic polymers include hexoses and 6-deoxyhexoses, pentoses, sugar alcohols, and ethers of any of these units. Gum arabic contains galactose, L-arabinose, L-rhamnose, and D-glucuronic acid in the ratio of 4:2:1:1. Each residue in the 1,3-linked galactose backbone is a branch point, with two different branch chains attached via 6,1-linkage on alternate galactose residues:

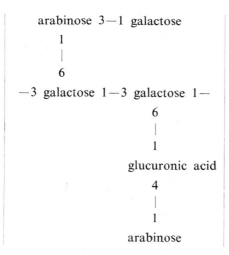

```
arabinose 3—1 galactose
    1
    |
    6
—3 galactose 1—3 galactose 1—
                6
                |
                1
          glucuronic acid
                4
                |
                1
            arabinose
```

Arabic acid and its salts have molecular weights of 250,000–300,000 and a diameter of about 100 Å. Gum tragacanth contains 1,4-linked D-galacturonic acid, L-fucose and xylose; the molecular weight is about 840,000. The molecule is elongate with dimensions of 4500×19 Å. Although there are many variations in the structures, sizes and specific monomers, these polymers fall, in principle, into a common category with the complex polysaccharides which have already been discussed.

Before leaving the wall polysaccharides, it would be well to summarize in a broad way their salient features as a group of natural structural materials.

In the construction of polysaccharides, it is clear that most of the possible ways in which any two monomers can be linked together have been employed, albeit to varying degrees. The specific monomeric units themselves are not great in number, although numerous enough to call for great skill and knowledge in the development of analytical techniques. Thus, the great array of polymers which are to be found in nature exist as a result of biosynthetic versatility. The combinations of monomer types and linkages, the degree of polymerization and the branching of the polymer together must account for the variety of wall polysaccharides which are now known. By virtue of its unusual size and high degree of molecular order, cellulose seems to be set off from all of the other wall substances as the primary frame substance. That the framework should be embedded in or encrusted with other polymers is not surprising, but the functional significance of the diversity which we have observed among the non-cellulosic polymers, and of their unequal, and sometimes unique distribution among various species or their parts remains a matter of physiological investigation.

Lignin

The cell wall polysaccharides are of interest as a specific part of cellular structure. Of even greater importance historically is the "Baustein" concept which has led in the past century to modern concepts of polymer structure both in biological and purely chemical pursuits. The polysaccharides together with the proteins and nucleic acids have provided the basic picture of macromolecules composed of repeating units, periodicity in organic molecules, and the coexistence of structural diversity with uniformity of overall pattern.

Relative to the polysaccharides and other biopolymers of importance, the lignins stand out as the unique aromatic polymers of the higher plant. They are also distinguished in their mode of synthesis, as will be discussed later, and in the manner in which monomers are joined together.

There is good reason to believe that our current picture of the cell wall polysaccharides as physical entities is moderately accurate at least. Careful isolation of cellulose and other carbohydrates and controlled solvent treatment and hydrolysis have sufficed to provide reasonably complete analytical data about these polymers. In contrast, the methods of historical importance when applied to the lignins are far more likely to modify the molecule and its monomers extensively.

The successful application of hydrolytic procedures to the polysaccharides has no true parallel among the lignins inasmuch as the linkage of monomers involves carbon-to-carbon bonds, rather than readily cleaved acetal or glycosidic oxygen bridges. Further, the phenolic and benzylic character of the lignin monomers renders them susceptible to undesirable oxidations during isolation and subsequent handling.

The lignins of some tissues dissolve to a limited extent in alcohols and other neutral organic solvents, but frequently form colloidal sols rather than true solutions. Dioxane is exceptional among organic liquids for its ability to dissolve the lignins; however, there is some question as to its inertness as a solvent. With dioxane as a possible exception, lignin extractants are commonly reaction solvents. Lignins are generally extractable with sulfites, bisulfites, alkalis, and acidic alcohols (for example, by alcoholysis), but are modified chemically by such processing. Lignin is resistant to water, dilute acids and cold concentrated acids.

The degradation accompanying isolation of soluble lignins has prevented extensive development of their detailed physical picture. Some concept at least of the molecule can be obtained from the study of these lignin derivatives in solution.

By isothermal distillation, lignins isolated from spruce and other woods yield molecular weights of 800–1000. From osmotic pressure measurements, molecular weights as high as 3800–4500 have been obtained. Various diffusion and sedimentation measurements have provided MW values of 800–12,000.

Recent molecular weight determinations carried out on lignin liberated enzymatically from pine, oak, and sugar cane yielded values of 695–890.

From the molecular weights of degradation (ethanolysis) products derived from synthetic lignin (from eugenol), minimum values of 1300–1500 were calculated.

Chemical molecular weight determinations based upon introduction of bromine, O-methyl groups and other substituents give values of 800–900, hence agree with many of the physical methods.

Clearly, we do not have certain knowledge about the molecular weight of the native polymers. In all probability, the lignins are quite polydisperse, hence give a variety of weights rather than one or a few. The native lignins probably exist in a wide range of molecular weights and sizes. It is difficult to distinguish between basic polymer units containing primary linkages and aggregates of these involving secondary forces. Viscosity and other physical measurements suggest more or less spherical molecules. The lower molecular units indicated by weights on the order of 1000 are unbranched, and have a $DP = 5$–6. The lignins exhibit a high index of refraction and strong aromatic absorption. Commonly, ultraviolet maxima center about 280 mμ, and range from 274–285 mμ. Rarely, lignins have strong absorption bands beyond 300 mμ. Extinction values (as $E_{1\ cm}^{1\ per\ cent}$) of about 100–200 have been found for pine, oak, birch, and maple lignins, and for synthetic lignin formed from eugenol in pea and celery tissue.

Lignins also possess identifiable infrared absorption bands. A band at 3400 cm^{-1} (about 2.9μ) is due to bonded OH groups. In the 1660–1700 cm^{-1} region, aldehydic or ketonic C=O appears; absorption bands at 1600 and 1510 cm^{-1} have been assigned to the phenyl skeleton whereas a 1430 cm^{-1} band is associated with aliphatic groups.

The electrophoretic behavior of the lignins has been little studied. The information now available suggests that some lignins (oak, for example) are quite homogeneous, whereas others (Scots pine, for example) give evidence of a multicomponent nature even after repeated purification. Pine lignin which contains one phenolic group is about 20 per cent higher in electrophoretic mobility than oak lignin, which contains three such groups. The negative charge on lignins must be due to dissociable phenolic or enolic hydroxyl groups. Electrophoretic measurements at pH 10 would allow some dissociation of these weak acidic groups. Phenols in general have pK values of 10–12, hence may be appreciably ionized at pH 10. The substituents on the benzene rings of lignin may further modify the pK.

Physical analysis by X-ray diffraction and polarized light demonstrates that isolated lignins lack the highly ordered crystalline structure found in many polysaccharides. The data agree with viscosity measurements which indicate a spherical form, and suggest a molecule with structural members arrayed so as to form a three-dimensional polymer, perhaps a reticulum. Although isolated lignins are optically symmetrical, they can exhibit form double refraction when carefully divested of the polysaccharide framework of the wall.

As we have already noted, the constitution of the lignin molecule has been studied by rather vigorous means which yield an imposing list of degradation products. Before we examine these substances, it would be well to examine the overall constitution of the lignins. It will be recalled that lignin may constitute one-half, or even more of the cell wall substance in older tissues, or may be indetectable (or absent) in youthful tissues. Commonly, it makes up 10–30 per cent of the weight of the cell wall. When this wall fraction is isolated, it commonly analyzes approximately 60 per cent carbon, 6 per cent hydrogen, and 15–20 per cent methyoxyl. The mode of isolation affects the constitution but a variety of methods may nevertheless yield similar data. In the case of spruce wood, "native" lignin isolated in small amounts with ethanol, and lignins isolated by the use of fuming HCl, 72 per cent sulfuric acid or cuprammonium solution yield, in the aggregate, the range C 62.4–63.6 per cent, H 5.3–6.4 per cent, OCH_3 14.6–16.0 per cent. Solvolytic agents which introduce alkoxy groups or modify the molecule otherwise will of course alter the constitution, but often in a predictable manner amenable to correction when actual percentages are calculated.

When lignins are subjected to pyrolysis, eugenol, *iso*eugenol, *n*-propylguaiacol, *o*- and *p*-cresol, guaiacol and catechol are included among the representative products.

Hydrogenolysis and other reductive processes tend to yield phenylpropanes with various side chain oxygen functions.

Alkali fusion yields acetic, oxalic, and protocatechuic (3,4-dihydrobenzoic) acids.

From methylated lignin treated with alkali and subsequently permanganate, veratric (3,4-dimethoxybenzoic) acid, isohemipinic (3,4-dimethoxy-5-carboxybenzoic) acid, and gallic (3,4,5-trihydroxybenzoic) acid are obtained.

One of the well studied degradative procedures is alkaline nitrobenzene oxidation. This treatment yields vanillin (3-methoxy-4-hydroxybenzaldehyde) from sprucewood. The corresponding acid, vanillic acid, and its 5-carboxylic derivatives are produced by oxidation of the aldehyde. Other aldehydes formed during nitrobenzene oxidation are *p*-hydroxybenzaldehyde and syringaldehyde (3,5-dimethoxy-4-hydroxybenzaldehyde). The latter occurs only in angiosperm lignins, not in conifers.

Mercaptolysis with thioglycolic acid and sulfonation of lignins yields solubilized acidic products containing, respectively the $-S-CH_2-COOH$ and $-SO_3H$ moieties. On mercaptolysis, each mole of thioglycolate taken up is accompanied by the appearance of a phenolic hydroxyl. Sulfonation, although more complex includes a similar phenomenon.

Other oxidation products of lignins include acidic benzene derivatives with 1,2,4,5-tetracarboxylic, pentacarboxylic, or hexacarboxylic (mellitic) acid patterns.

The aldehydes formed on nitrobenzene oxidation vary considerably in amount from species to species. Hydroxybenzaldehyde ranges from less than 1 per cent to as much as 10 per cent of the total lignin and vanillin from 17–25 per cent. The syringaldehyde content of hardwoods ranges from 2–13 per cent. These aldehydes may comprise *in toto* as much as 40 per cent of the overall weight of lignin samples, or as little as 20 per cent.

The information carried by this mass of analytical data is truly impressive. It is without doubt a key to many of the problems of lignin structure including the nature of the aromatic groups and their hydroxylation and methoxylation patterns; the nature of side-chains and their oxygen functions; and the nature of the linkage between monomers.

Cross and Bevan, and others, in past years, sought to account for the results of degradation studies by invoking a structure in which these degradation products are derived from a uniquely constituted molecule. As more complete analytical data were accumulated, both the inadequacies of such unique structures and the utility of the "Baustein" concept became apparent. Although structure among the lignins is far from being resolved in detail, a representative picture can now be put forth. The idealized structure of a conifer lignin which is illustrated below does not represent any one polymer segment, but rather attempts to incorporate in one structure most of the salient features of these substances.

For the guaiacylpropane heptamer depicted we may calculate a molecular weight of 1272, and the composition, C 65.1 per cent, H 5.9 per cent, OCH_3 17.1 per cent.

The molecular formula $C_{69}H_{75}O_{23}$ when normalized to the phenylpropane building stone would yield $C_{10}H_{10.8}O_{3.3}$. By comparison, the formula for coniferylalcohol, one of the important precursors of lignin is $C_{10}H_{12}O_3$. Thus, the theoretical polymer segment corresponds to a multiple of a slightly oxidized coniferylalcohol unit. Such a structure does not account for all known degradation products, but can explain the majority of them. Notable for its absence is the pyrogallol dimethyl ether (syringyl) groups found in angiosperms.

The chemical groups which join the benzene rings in this model structure are: (a) the prevalent glyceryl-β-phenyl ether linkage at rings 1→2, 3→4, 6→7; (b) the phenylcoumaran, or hydrofuran, linkage which occurs at rings 2→3; (c) the pinoresinol or lignane linkage, which occurs at rings 4→5; and (d) the biphenyl linkage which occurs at rings 5→6. Thus, both ether bridges and carbon-to-carbon bonds are present in side chain-to-nucleus, side chain-to-side chain, and nucleus-to-nucleus condensation. The common hydroxylation–methoxylation pattern is that of guaiacol, which accounts for the vanillin formed by oxidation. Primary and secondary alcohol groups and the less frequent free phenolic group are accounted for, and provision is made for branching via the open oxy-group on ring 6. The phenolic OH at ring 5 might provide an alternative branch point, but stereochemical limits would restrict branching to one of these rings only. The end monomer unit (ring 7)

Representation of a theoretical guaiacyl propane heptamer. After Adler, E., *Newer Views of Lignin Formation*. No. 40, p. 294. Technical Association of Pulp and Paper Industries, Easton, Pennsylvania, 1957.)

corresponds to coniferaldehyde. The unsaturated aldehyde group by undergoing condensation with phenols, amines, and the like forms the chromogens which account for some lignin color reactions. Other aldehydic groups may replace primary alcohols, but not extensively.

Among the variations of this pattern, the closure of the open glyceryl-β-phenyl ether chains to form additional ring structures as shown in the 2→3 linkage is not unlikely. Some lignins may be more homogeneous, consisting, for example, of repeating units containing the 2→3 linkage. Others are more diversified in monomer types.

As the main guide lines to an understanding of lignin structure become established, more attention is directed toward the organic and physical chemistry of the cellular synthetic processes which lead to the formation of such unusual yet widespread biopolymers. These aspects of lignin research will be examined when we consider some aspects of cell wall dynamics and surface chemistry.

Important Minor Wall Components

The ash constituents of cell walls commonly make up 1–5 per cent of their dry weight. Many elements in varying quantities are to be found among the mineral substances of the wall, including calcium, magnesium, phosphorus, and silicon. In some plant groups relatively high inorganic residues containing considerable proportions of SiO_2 are to be found. Sorghum, wheat, and maize are representative of the Graminae which contain 10–12 per cent ash, principally silica. Sorghum and wheat are extreme cases, with inorganic residues running 90 per cent SiO_2. Among dicotyledonous plants, sunflower and *Lantana* contain about 11 per cent ash of which onequarter is SiO_2. *Equisetum* is notable among the non-flowering vascular plants with its silica content of 8–10 per cent.

The study of mineral (ash) constituents of cells involves the technique of microincineration. In general, longitudinal or transverse sections, or whole tissue if sufficiently thin are pressed between microscope slides and ashed at 450–900°C. The upper slide is then removed, Canada balsam added, and the cleared material placed under a standard cover glass. The resulting mounted residue from incinerated tissue is known as a spodogram.

By this procedure, mineral substances are frequently retained in their natural position in the cell. In spite of the dimensional changes which may occur, relative locations (for example, nuclear vs. cell wall residues) may be distinguished. Both microchemical and microphysical methods, including petrographic and crystallographic analysis may be applied to these inorganic components of tissues, cells, or parts of cells.

The siliceous residue may vary considerably in particle size and shape. In the sorghum leaf, irregularly shaped particles measuring approximately 0.007×0.0152 mm; dumb-bell shaped particles measuring 0.0135×0.0207 mm; and rectangular particles 0.017–0.073 mm long may be found. The leaf sheath in wheat contains oval particles 0.017×0.0153 mm.

In bamboo, the nodes contain particles 0.41–0.57 mm in length. The deposition of silica in *Lantana* is restricted to trichomes and epidermal cell walls in the lamina. The silicified trichome measures at maximum, about 0.5 mm in length. The irregular particles in the sunflower have a maximum diameter of about 0.06 mm. The ash of sorghum, wheat, maize, sunflower and bamboo contain an isotropic silica with an index of refraction of 1.45.

In *Lantana* most of the silica is like that in the other species, but also contains a form of silica showing positive double refraction ($\omega = 1.54$, $\varepsilon = 1.55$). The crystallographic character of the two forms of silica has been established by X-ray diffraction analysis. The common type of pattern is that of the amorphous SiO_2, opal, and is identical with geological deposits of diatom (biogenetic) opal. The second component of the *Lantana* trichome gives the X-ray diffraction pattern of α-quartz. Petrographic and X-ray diffraction analysis are in complete agreement.

Some comparative aspects of cell wall mineral substance are yet to be considered, but it should be noted that a great deal remains to be understood about the chemistry, physics, and physiological significance of inorganic wall components.

Although lipids are often present only in small amounts, they are nonetheless important as a cell wall component. Their hydrophobic character is associated with water-proofing and perhaps other protective functions. The lipophilic wall substances associated with the cuticle constitute a quite diverse group from the structural standpoint. Their location and the common properties which are conferred by the long hydrocarbon chains and low proportions of polar functions serve to place them together.

Cuticular substances may be located by their affinity for lipid dyes or basic lipid dyes. Micromelting tests may be used in conjunction with optical analysis in the differentation of cuticular substances.

Chemically, the cuticle consists of substances soluble in benzene, ether, pyridine and the like, together with more refractory materials which may be solubilized only by saponification. The extractable cuticular substance has sometimes been designated collectively as "waxes". These substances actually consist of paraffins, aliphatic acids, aliphatic alcohols and the monesters which are, properly, the waxes. Both the alcohols and acids consist of the even-numbered members of series lying between C_{16} and C_{34}. The paraffins consist of odd-numbered unbranched members of the series C_{27} to C_{31}. The saturated acidic components include palmitic $C_{15}H_{31}COOH$, stearic, $C_{17}H_{35}COOH$, arachic $C_{19}H_{39}COOH$, cerotic $C_{25}H_{51}COOH$ acids, and others up to $C_{33}H_{37}COOH$. The unsaturated acids are represented by oleic acid, $C_{17}H_{33}COOH$, and linoleic acid $C_{17}H_{31}COOH$. The important fatty alcohols are cetyl C_{26}, myricyl C_{30}, and higher homologs up to C_{34}. Cerin, $C_{30}H_{50}O_2$, and friedelin, $C_{46}H_{76}O_2$, are cyclic alcohols containing an alcoholic OH and a cyclic ether bridge. In all of these aliphatic derivatives, there is but one functional group per molecule. Hence, the formation of waxes by esterification can involve only one molecule each of acid and alcohol. In addition, there are esters which yield on saponification by bi-functional ω-hydroxy acids such as hydroxy-lauric (or sabinic) and hydroxy-palmitic (or juniperic) acid. These acids are joined "head to tail" to form the high molecular weight polyesters known as "estolids". In general, the degree of polymerization in the estolids is not too high, as indicated by their melting ranges and solubilities.

The most conspicuous sources of the simpler cuticular substances are leaf epidermis, fruit epidermis, and seed coats. Carnauba wax, from the leaf of the palm, *Copernicia,* contains esters of the even C_{26}-C_{34} alcohols with monocarboxylic acids of the same chain lengths. The paraffin $C_{27}H_{56}$ also occurs. Candellilla (*Euphorbia*) wax consists of 50–60 per cent *n*-hentriacontane ($C_{31}H_{64}$), 15 per cent C_{30}, C_{32}, C_{34} acids, and 5 per cent of the corresponding alcohols. Apple cuticle contains *n*-$C_{27}H_{56}$ and *n*-$C_{29}H_{60}$, together with the C_{26}, C_{28} and C_{30} alcohols.

The resistant cuticular substances are complex polymers whose structural plan is still largely unknown. Three high-polymeric cuticular substances have been recognized: suberin, which is associated with cork or "corky" tissues; cutin, which is the common cuticular ground substance present on epidermal cells in general; and sporopollenin, the extremely resistant wall substance of pollen grains. Sporopollenin is the most highly polymerized, most difficultly saponifiable of the three, suberin is the most readily decomposed.

Oxidative degradation of suberin yields a variety of aliphatic acids, among them suberic acid, $HOOC(CH_2)_6COOH$; phloionolic acid, $C_{17}H_{32}(OH)_3COOH$; phloionic acid, $HOOCC_{16}H_{30}(OH)_2$ $COOH$; phellonic acid, $C_{21}H_{42}(OH)COOH$; and eicosane dicarboxylic acid; $HOOC(CH_2)_{20}COOH$. The presence of hydroxyacids, and dicarboxylic acids which may have been formed from hydroxyacids, suggests that suberin, like the estolids, contains poly-ester. In addition ether linkages have been indicated. Some of the acid groups are, apparently free. This is shown in cutin by anionic character and an affinity for basic dyes. The polyfunction-ality of several suberin derivatives could lead to a cross-linked tri--dimensional reticular structure. Such a proposal structure is borne out by optical analysis, which shows that the highly polmerized cuticular network is isotropic.

III. The Wall as a Unit

The properties of cellulose have been discussed from the view-points of polymer chemistry and crystallography. Further X-ray examination of fiber walls indicates that they are not uniformly crystalline in character. Rather, the wall seems to contain cellulose blocks measuring 55 Å in cross-sections by 600 Å in length. The indicated length corresponds to $600/10 = 60$ cellobiose units, approximately, or 120 anhydroglucose residues. In the wall there-fore, the DP of 10^2 indicated above must be compared with mole-cular weight determinations yielding $DP = 10^3-10^5$, with the most careful measurements favoring the range 10^4-10^5. From these comparisons it is evident that individual, extended molecular chains are so organized that they participate in many crystallites, fray-ing out in the intercrystalline regions.

From cellobiose to the organized wall glucopyranose chains are associated with one another in definite and orderly patterns. Individual chain molecules measuring 33 Å^2 in cross-section aggregate into *elementary fibrils*, 3000 Å^2 in area, each such fibril containing 100 glucopyranose chains. Twenty elementary fibrils form a *microfibril* which is 250 Å on a side and 62,500 Å^2 in area. Individual chains are amicroscopic whereas elementary and micro-fibrillar structures are submicroscopic. The aggregation of some 250 microfibrils forms the microscopic *fibril* 0.16 μ^2 in cross--section. Finally, some 1500 fibrils are organized into macroscopic *fiber* such as the cotton hair over 300 Å^2 in diameter. Thus, a single fiber must contain in cross-section some 7.5×10^8 glucopyranose chains.

The specific gravities of dry fibers approximate the calculated density of crystallite regions (1.60±0.05), hence they must contain relatively little vacant space. Upon hydration, inter- and intra-fibrillar spaces increase appreciably, finally becoming a system of fluid-filled cavities. Within the microfibrils, these spaces average 10 Å in width, but the intermicrofibrillar spaces may be as much as 84 Å wide.

The fundamental ordered structural element of this fibrous system is the elementary microfibril, or micelle. This is the most perfectly ordered unit, 50–70 Å wide and at least 600 Å long. This unit then corresponds to the crystalline region whose presence is indicated in the wall. The paracrystalline cellulose in the inter-micellar spaces contains molecular chains oriented parallel to the fibrillar axis but lacking complete three-dimensional intramicellar order (Fig. 1).

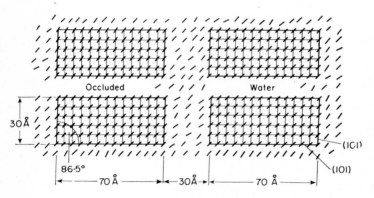

FIG. 1.   The micellar structure of cellulose. (From Wardrop, A.B. and Bland, D.E., The process of lignification in woody plants. *Proc. 4th Intl. Congress of Biochem.*, vol. II, p. 93, 1959.)

Although some fiber cell walls, ramie, for example, consist of cellulose chains oriented almost perfectly parallel to the fiber axis, this condition is not common or typical.

More common than the true fiber structure is the fiber-like texture in which orientation of molecular chains parallel to the fiber axis is disturbed by random scattering. The optical properties of such walls result from the statistical arrangement of chains along the fiber axis. Of widespread occurrence, especially in secondary walls, is the spiral texture. The angular deviation of spiralled chains from the fiber axis is described as their angle of ascent. In hemp, the outer layer of the secondary wall ascends to the left at 28° whereas the inner layer ascends at only 2°. In cotton, the inner layer of the secondary wall ascends at 24°, the outer at 35°, with reversals in the direction of spiral. Even ramie has a slight angle of ascent. In the cotton fiber and secondary walls of tracheids, a lamellated annular structure may occur. In contrast to the textures already examined, this so-called ring structure is produced by parallel chains lying perpendicular to the fiber axis. A variant on ring structure which occurs very commonly in many primary walls is the tube structure. The cellulose chains exhibit considerable scattering, but resemble the ring structure in their general orientation.

The full extent of cellulose organization in the cell wall is best shown in wood fiber cells (Fig. 2), which possess a tubular primary wall and a tripartite secondary wall containing an outer double spiral, a middle steep spiral in which fibrils are arranged in concentric lamellae, and an inner flattened spiral.

The overall relationship among the variously oriented fibrils approaches orthogonality. This condition is to be found in another great bio-mechanical system, bone, in which organo-mineral lamellae laid down along compression and tension lines are arranged orthogonally.

Within the reticulum formed by organized cellulose fibrils and microfibrils is located the whole array of non-cellulosic substances. These extend beyond the primary wall itself as an intercellular layer of amorphous character partially identified with the term, middle lamella. In young cells, the intercellular gel is commonly rich in calcium and magnesium pectates, but may also contain pentosans and dispersed cellulose fibrils. The non-cellulosic polysaccharides which extend into the primary wall include pectic substances and cellulosans. In the secondary wall, cellulose predominates, but it has been claimed that the hemicelluloses present serve as binding agents for cellulose aggregates or bodies.

PLATE II

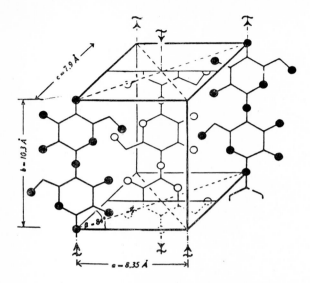

Unit cell of cellulose (after Meyer and Misch)

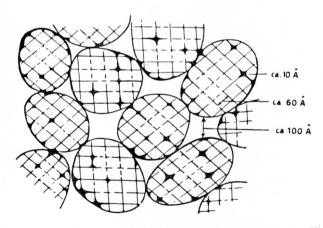

Microfibrils from a Fibril Showing Diagrammatically
their Elementary Fibrils (after Frey-Wyssling)

FIG. 1.

PLATE II

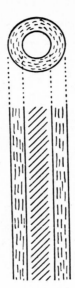

Fibre Structure
of Ramie

Spiral Structure
of Cotton

Tubular Structure of
Young Primary Walls

Fibrillar Orientation in Three Cell Wall Types

FIG. 2.

With the onset of maturation the cell wall commonly becomes encrusted with lignin. The extent of lignification varies, but in the main it is a process associated with intercellular substance and the primary wall. In spruce wood, for example, 60–90 per cent

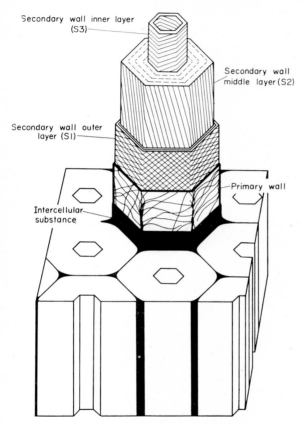

FIG. 2. Texture and lamellar arrangement in the cellulose frame of wood fiber cells. (From Wardrop, A.B. and Bland, D.E., The process of lignification in woody plants. *Proc. 4th Intl. Congress of Biochem.*, vol. II, p. 96, 1959.)

of the lignin is present in the intercellular layer, but only 10–12 per cent occurs in the secondary wall. In other instances the inter-cellular and primary layers have been reported to consist of 71 per cent lignin, 14 per cent pentosan, and 4 per cent cellulose. The deposition of a carbohydrate wall generally precedes lignification.

It has been claimed that lignin and the polyuronides are closely associated and that cellulose is aggregated with xylans or mannans. On the other hand the chemical combination of xylose and xylobiose with lignin has been demonstrated in beechwood, where there is no evidence for a lignin–uronide linkage.

The existence of chemical as well as physical associations between lignins and wall carbohydrates seems well established even though the nature of the chemical linkage may vary. The intimacy of the association between lignin and other interstitial substances whether physical or chemical is further shown by the diminished affinity of lignified tissues for various stains. Reduced staining with ruthenium red indicates the masking of or combination with polymeric acid groups and the reduced affinity for congo red has been interpreted as evidence for the infiltration of lignin into intermicrofibrillar spaces. Finally, metal particles deposited in lignified fibers are smaller than those deposited in delignified cell walls.

From the foregoing considerations we may envisage a framework of oriented cellulose fibrils which form in depth a crossed (orthogonal) system of considerable strength. This frame is then embedded in a gelatinous matrix composed of generally lower molecular weight polysaccharides. In maturing tissues, the polysaccharide matrix and lignins interpenetrate, and, in many instances, the latter becomes the dominant interstitial polymer.

The major constituents to be found in various regions of the cell wall differ primarily in their proportions, rather than in the unique presence or complete absence. Thus, cellulose varies from a dispersed fibrillar condition in the intercellular layer to a dense reticulum in the inner region of the secondary wall. The interstitial substances accordingly show a reciprocal variation. Both the intercellular and primary layers contain open structures which could conceivably be deformed considerably by compression. Although the compact interstitial gel, unlike cellulose, can have little tensile strength it may exhibit more resistance to compression. Hence, the combination of frame and matrix is reminiscent of a reinforced concrete wherein comparatively slender and flexible steel rods of great tensile strength are buttressed by an amorphous imbedding medium which has little resistance to extension forces, but great resistance to compression.

Among the imbedding substances, lignin is distinctive in many ways, as we have seen. One physical property, a consequence of a predominatly hydrocarbon constitution is its hydrophobic character. Cellulose undergoes moderate to appreciable swelling and shrinkage as its hydration water is varied. The interstitial polysaccharides are less ordered than cellulose and can become more hydrated. Consequently, these substances must undergo extreme volume changes according to their state of hydration.

The incrustation and interpolation of hydrophobic lignins into the hydrophilic gel and onto the surfaces of cellulose microfibrils introduces a structural element of comparatively constant volume into the reticulum thereby providing the cell wall with a dimensional stability to a degree otherwise impossible. This property may be of greatest moment relative to the loss of water from plant tissues during maturation and senescence. The effect of lignins on staining, on interstitial free space and shrinkability and its marked form double refraction all support the concept that it provides a stable element of bulk in the wall.

Although lignins are, quantitatively, the major hydrophobic substances of the cell wall, young unlignified tissues and non-mechanical superficial cells may contain cuticular substances or other lipoidal materials.

In the young primary wall with tubular texture, X-ray interferences corresponding to periods of 60 and 83 Å have been identified with waxes intercalated into the microfibrillar system. Rod--shaped wax molecules are oriented at right angles to the micellar strands of the wall. As in the case of lignin deposition, these extremely hydrophobic molecules mask the polysaccharides, particularly the individual cellulose strands. In cutinized walls, the outermost material is cutin alone, but most of the cuticular layer consists of hydrophilic lamellae of cellulose and pectin, layers of radially arranged wax, and between them, randomly arranged cutin. Cutin possesses both hydrophilic and hydrophobic groups, hence can serve as a "cement" holding together the hydrophilic polysaccharides and hydrophobic waxes. In succession, we have, from the cuticle inward: cutin only, cutin, plus wax, cutin–wax–polysaccharide, and, innermost, if present at all, typical polysaccharide layers.

Lignin as an internal substance serves as a mechanical and water-proofing agent. Cuticular substances retard water loss, and possess ultraviolet absorbing constituents which may act as a superficial radiation screen.

PLATE III

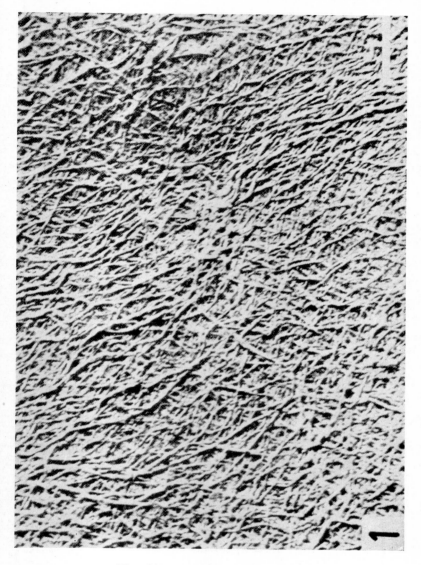

Fine Structure of Plant-Cell Walls.
FIG. 1. Primary cell wall; cambial cell of *Fraxinus excelsior.*

PLATE III

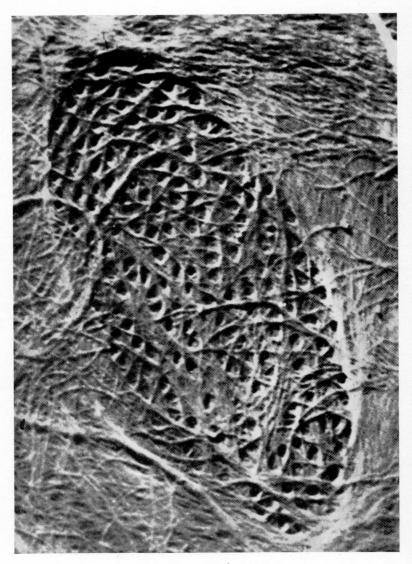

Fine Structure of Plant-Cell Walls.
FIG. 2. Pit membrane of a simple pit; root of *Zea Mays*.

CHAPTER 2

CELL WALL DYNAMICS

I. Biosynthesis, Growth and Regulation

Commonly, the cell wall is pictured as something of a final repository for certain products of cellular metabolism. Thus, we see cell wall rudiments laid down during division, increase in extent during growth, and finally undergo thickening and often incrustation with secondary substances such as lignin, suberin, or cutin. Even though the cell wall, once deposited, tends to retain its constituents, its relations to growth and differentiation are not those of a static member in living cells and tissues. The apparently static character of cell walls probably derives from the fact that they commonly outlast the living protoplast.

During the active life of the cell, the changing pattern of cell wall constitution and organization is a reflection of the changing physiological states and biochemical capacities of the protoplast. Thus, the chemical and physical properties of cell walls provide a measure of protoplast and membrane functions which we now comprehend but vaguely.

Our present consideration of cell wall dynamics will center about three areas: First, current knowledge as to the biosynthesis of individual cell wall components; second, the patterns of change in constitution and architecture during growth and differentiation; and third, the little knowledge now at hand about chemical regulation of cell wall formation.

The Biosynthesis of Cell Wall Components

The overall composition of the cell wall changes during growth. Thus, a sunflower hypocotyl of 5 mm diameter contains 38 per cent cellulose, 46 per cent pectin, and about 8 per cent each of (neutral) non-cellulosic polysaccharide and lignin. A five-fold increase in diameter is accompanied by marked increase in neutral polysaccharide and lignin, and slight increase in cellulose, and a marked decline in pectic substance. The 25 mm hypocotyl contains about 42 per cent cellulose, 14 per cent pectin, 24 per cent neutral polysaccharide, and 20 per cent lignin. Analysis of woody species shows similar changes. For example, in ashwood, the cambium contains about 22 per cent cellulose, 31 per cent pectin, and 6 per cent lignin. In new wood, these components run 55 per cent, 4 per cent and 25 per cent, respectively, and in sapwood, 59 per cent, 2 per cent, and 21 per cent. These examples show that the proportions of cell wall components change markedly during growth and maturation, but they can be assigned no other significance. The apparent decrease in pectic substances which occurs as lignin and other constituents increase demonstrates only that the synthesis of lignin has become a dominant process and that pectin synthesis has become slower, or has perhaps ceased. Although such data show no biogenetic relations, they are nevertheless useful. A change in proportions among wall components is important in the study of developmental chemistry and reflects the basic character of growth and maturation processes in plant cells.

Remarkably little is known about the biochemistry of wall synthesis in vascular plants. General biochemical information about polysaccharide synthesis can be drawn upon for analogies; however, such excursions do not provide specific information about the relation of wall synthesis to general cell economy. In contrast to polymers of α-glucose such as amylose whose biochemistry has been moderately well known for some years, information pertaining to the enzymology of cellulose synthesis has been acquired only recently. Indeed, the only definitive information has been derived from studies of the micro-organism *Acetobacter xylinum* which is notable as a cellulose producer. When C^{14}-labeled carbohydrate intermediates are supplied to *Acetobacter* the label appears in cellulose when provided by D-glucose-1-C^{14}, D-glucose-2-C^{14}, D-glucose-6-C^{14}, D-mannitol-1-C^{14}, or glycerol-1-3-C^4 the sole carbon source. Although ethanol enhances incorporation of labeled glucose in cellulose, labeled ethanol and acetate are not themselves incorporated when provided either as sole carbon sources or in the presence of glucose. When glucose-1-C^{14} is supplied exclusively, some label appears at the 3- and 4-positions. Although most of the hexose supplied appears to undergo direct conversion to cellulose, therefore, a significant proportion first undergoes cleavage. Following evidence that cellulose formation by *Acetobacter* is extracellular, it has recently been shown that a cell-free enzyme system for the direct polymerization of glucose does indeed exist in this organism. This system is characterized by a pH optimum of 8.5–9 and an absolute requirement for adenosine triphosphate. It is entirely possible that cellulose synthesis in vascular plants involves different enzymes and mechanisms, and, indeed, it seems that intact *Acetobacter* itself has a secondary indirect mechanism. Nevertheless, the relative simplicity and directions of this cell-free system are appealing, and its demonstration is of great significance. It has been established, moreover, that glucose-1-C^{14} is a good precursor of cellulose in wheat, the label appearing mainly at C_1.

When glucose-1-C^{14} is fed to wheat and other plants, the C_1 label appears not only in cellulose, but also in xylan, D-xylose, L-arabinose, and galacturonic acid. Labeled xylose and ribose are poor and indirect precursors of xylan, whereas glucuronolactone-1-C^{14}, a poor cellulose precursor, is readily converted into xylan. Current evidence supports a pathway to hemicelluloses involving oxidation at the C_6 position in glucose to form intermediate uronic acid with subsequent decarboxylation to pentose. In contrast, xylose is first converted to glucose-6-phosphate via *trans*-ketolase and *trans*-aldolase and then transformed into xylan by C_6-decarboxylation.

The concept of C_6-decarboxylation is relatively novel, and counter to the somewhat earlier picture which emphasized reactions at the glucose C_1. Even earlier, it was believed that hemicellulose arose from cellulose by oxidation to polyuronide and decarboxylation to xylan. Although there were many appealing features in this decarboxylation theory, it became untenable as more detailed information about pentosan structure was obtained. The modified decarboxylation theory which has been recounted, centers about monosaccharides rather than polysaccharides. The most promising mechanism which has been put forth to account for the close interrelationships among polysaccharide precursors is described in the nucleotide theory of polysaccharide synthesis. According to this concept, uridine disphosphate-glucose (UDPG) and similar nucleotide sugars are the pivotal intermediates in a pool of glucosyl donors for polysaccharide synthesis.

UDPG and UDP-D-galactose are readily interconverted. UDPG can be oxidized to its glucuronic acid. UDP-N-acetylglucosamine is known, and has been proposed as the chitin precursor in fungi. Similarly, guanosine diphosphate mannose is known in yeast where it is presumed to serve as the mannan precursor. UDPG itself is believed to be involved in the synthesis of cellulose in *Acetobacter*. Uridine diphosphate glycosides in vascular plants include the D-glucose, D-galactose, L-arabinose, D-xylose, and D-glucuronic acid.

Thus, interconversions among the free sugar pool, the phosphate ester pool and the nucleotide glycoside pool, and the mechanisms for transfer of glycosyl groups offer the outlines for a fuller understanding of wall polysaccharide synthesis in the near future.

The chemical uniqueness of lignins among wall polymers allows them to be distinguished from the much intergraded wall polysaccharides, an advantageous property. Nevertheless, although the outlines of lignin synthesis are clear, the broad physiological and biochemical details of synthesis remain to be established.

In the early days of biopolymer chemistry, a great deal of interest was focused upon the hydrolytic enzymes which were obtained from many sources with comparative ease. These enzymes were recognized as catalysts for the degradation of proteins and polysaccharides, but were also viewed as agents of significance in synthesis as well. Although equilibria in proteinase and carbohydrase systems are far over toward hydrolysis, some of these enzymes could be used for synthesis under special conditions. Accordingly, the early view that polymer synthesis was a reversed hydrolysis was difficulty sustained, but seemingly necessary. As we now realize, the synthesis of biopolymers can be accomplished in general only with the driving force of phosphorylative mechanisms and the mediation of nucleotides and other carriers.

In contrast, there has been a remarkable consistency in the very basic notions of lignin synthesis. The outstanding degradative studies by pioneers such as Klason and Freudenberg, and others, led early to the "Baustein" concept of lignin structure, and thereby provided some notions about lignin precursors. Again, chemical syntheses of simple models with lignin-like degradation patterns came comparatively early in the hands of Klason, Erdtman, Freudenberg, and others. In principle, the biochemistry of lignin synthesis as it is now understood is in large measure a confirmation of the organic chemical picture established before.

The study of lignin biosynthesis has been carried out in two types of systems. First, of course, are those procedures in which living tissue is provided, with the presumed lignin precursor and the appearance of polymer determined. Such "feeding" experiments are divisible, in turn, into long-term studies with labelled precursors and short-term studies with high concentrations of unlabelled precursors.

Second is the *in vitro* model which again has followed two subordinate lines, the purely enzymological and the enzyme–polysaccharide (matrix) system.

This classification of approaches omits histochemical procedures which bear upon the more physiological study of lignification as a differentiation phenomenon. But these approaches provide information about precursors, their proximity to lignin in the biogenetic scheme, the enzymes which are involved, and something of the conditions under which experimental lignin formation takes place.

The list of species used in experimental lignin synthesis has grown considerably in recent years. Woody species such as spruce, eucalyptus, bamboo, willow and maple and herbaceous plants including bean, pea, potato, elodea, celery, buckwheat, clover, wheat, and timothy.

Among the substances which have been established as precursors in various experiments are the amino acids phenylalanine, tyrosine, and dihydroxyphenylalanine (DOPA); phenols such as eugenol, *iso*eugenol, coniferyl alcohol, and sinapyl alcohol; the aldehydes, vanillin, *p*-hydroxybenzaldehyde, coniferaldehyde; caffeic, ferulic, and phenylpyruvic acids; and glucosides such as coniferin, syringin, and glucovanillin.

Enzymological studies have focused attention upon phenol-dehydrogenase mixtures from the conifer *Araucaria* and from common mushrooms; laccases from mushrooms, and the Japanese lac tree; and peroxidases from a great variety of tissues. The laccases and peroxidases are associated with the terminal steps in lignin synthesis, the oxidative polymerization reactions themselves. These reactions correspond to the general formulation:

$$C_6-C_3 \xrightarrow[\text{peroxidase}]{\text{laccase or}} (-C_6-C_3-)_n$$

phenylpropane   lignin
monomer   polymer

This formulation is too general, however, to provide an accurate picture of the transformations involved. Thus, a number of supporting or preparative reactions and enzymes must be taken into consideration. Among these reactions are:

(a) Glucoside hydrolysis, as in coniferin $\xrightarrow{\beta\text{-glucosidase}}$ coniferyl alcohol + glucose.

(b) Deamination as in tyrosine + α-ketoglutarate $\xrightarrow[\substack{\text{pyridoxal-5-}\\\text{phosphate}}]{\text{transaminase}}$

p-hydroxyphenylpyruvate + glutamate, and

tyrosine $\xrightarrow[\text{FAD}]{\text{amino acid oxidase}}$ p-hydroxyphenylpyruvate.

(c) Hydroxylation as in

phenylalanine $\xrightarrow{\text{hydroxylase}}$ tyrosine

tyrosine $\xrightarrow{\text{tyrosinase}}$ DOPA.

(d) Aldehyde reduction as in

$$\text{coniferaldehyde} + \text{DPNH} \xrightarrow[\text{dehydrogenase}]{\text{alcohol}} \text{coniferyl alcohol} + \text{DPN}^+.$$

(e) Transmethylation as in

$$\text{3,4-dihydroxyphenylpyruvic acid} \xrightarrow[\text{methyltransferase}]{\text{methionine}}$$

3-methoxy-4-hydroxyphenyl (guaiacyl) pyruvic acid.

(f) Side chain lengthening as in

$$\text{vanillin} + \text{acetaldehyde} \xrightarrow{\text{carboligase}} \text{guaiacylacetyl carbinol}$$

$$\text{vanillin} + \text{pyruvate} \xrightarrow{\text{carboxylase}} \text{guaiacylacetyl carbinol} + CO_2.$$

(g) Side chain oxidation as in

$$\textit{iso}\text{eugenol (eugenol)} \xrightarrow{O_2} \text{coniferaldehyde.}$$

Obviously, a number of the foregoing reactions must fall into appropriate sequential arrangements if certain potential precursors are to be converted into lignin. Thus, phenylalanine would be transformed into guaiacylpyruvic acid via hydroxylase, tyrosinase, transmethylation, and transaminase or amino acid oxidase. The successive intermediates here would be tyrosine, DOPA, and guaiacylalanine.

Glucovanillin would be subjected to β-glucosidase, carboligase (or carboxylase). and alcohol dehydrogenase to yield guaiacyl methyl ethylene glycol.

In short term, "forced" peroxidation with high levels of precursor and peroxide, eugenol is converted into lignin via coniferaldehyde. Coniferaldehyde or its alcohol are converted directly to lignins, and ferulic acid is polymerized directly to a low-aldehyde lignin. On the other hand, phenylalanine is inactive and tyrosine and DOPA are transformed into quinonoid polymers rather than lignin. Coniferin is converted to lignin only after a lag period while coniferyl alcohol is released from the glucoside.

Although lignin polymers can be formed by several oxidizing enzymes, peroxidases are the most widespread hence must figure most importantly in plants as a whole. In addition to slow polymerization by cell-free preparations, crude or partly purified enzymes, both the rate and specificity are subject to modification by the presence of polysaccharide surfaces (matrix effects discussed in Chapter 7).

We have now examined a number of aromatic precursors of lignin and their behavior, but have not given consideration to the obviously important pre-aromatic substances which are intermediates in the conversion of carbohydrates into phenylpropanes. Although it has long been appreciated that the C_7 sugar sedoheptulose played a part in the synthesis of aromatic substances, the main lines of biosynthesis have only recently received some clarification. The biosynthetic pathway proceeds from sedoheptulose-1,7-diphosphate through 2-keto-3-deoxy-7-phospho-D-glucoheptonic acid to a series of alicyclic derivatives. Of particular note are shikimic acid, a trihydroxylated cyclohexene-carboxylic acid and prephenic acid, 1-carboxy-4-hydroxy-cyclohexadienyl-pyruvic acid.

The biogenetic scheme provides for several aromatic compounds:

protocatechuic acid

↑

5-dehydroquinic acid→5-dehydroshikimic acid

↓

tyrosine←prephenic acid←shikimic acid

↓

phenyl pyruvic acid→phenylalanine

The biochemistry of cell wall synthesis is obviously far from complete, but the present state of knowledge is sufficient to indicate some of the major pathways and processes involved.

Changes in the Walls of Growing Cells

In the primary walls of the *Allium* rootcap, pentoses and hexoses are high whereas the hexuronic acid content is low. During the radial (isodiametric) enlargement, which is concurrent with cell division all three groups of carbohydrates are low, but increase proportionally to the increase in wall surface area. These carbohydrates increase markedly with the onset of cell elongation, their rate of synthesis exceeding in magnitude the increase in cell surface area.

The apical initials of the *Allium* root contain small amounts of pectin, protopectin and cellulose whether calculated on a per cell or wall area basis. Protopectin is relatively more abundant in the walls of apical initials than in radially enlarging cells. The proportion cellulose:pectic substance is the same in apical initials and cells in radial enlargement. In the transition between radial enlargement and elongation where both processes coincide, the open-work wall of the radial phase becomes more continuous, the gaps are filled in, and the primary pit fields are reduced in size. During this transition, all carbohydrates increase, and pectin, soluble non-cellulose polysaccharide, and cellulose all increase more than the surface area of the wall.

With the cessation of radial enlargement and onset of rapid elongation the discontinuities of the wall are filled in with cellulose, pectin, hemicellulose and non-cellulosic polysaccharide. Thus, these components increase per unit area immediately following the end of the radial phase and then become proportional to wall area at a new high level, although the wall as a whole is not thicker. It has been suggested that elongation may occur only when there is an approximately continuous wall, hence is limited during the radial phase by its fragmentary character.

In proliferating tobacco callus tissue, some 25 per cent of total cell nitrogen is associated with the insoluble cell wall fraction, whereas non-proliferating pith walls contain but 2 per cent of the cellular nitrogen. Wall protein is rich in hydroxyproline relatively abundant in histidine, and poor in thio-amino acids. Tyrosine-C^{14}, is rapidly incorporated into the wall protein, but its radioactivity remains fixed and is not redistributed throughout the various amino acids. It is evident that the tyrosine metabolism of the wall is of a terminal rather than intermediary character.

The introduction of methyl ester groups is one of the important changes in the pectic substances during wall development. The methyl carbon of partially esterified pectic acid (pectinic acid) can originate from a variety of sources, including methionine, formaldehyde, glycine, and serine. Of the C_1 compounds, formate is the best methyl source and serine the poorest. In contrast to the efficiency with which formate becomes ester methyl, it seems to be a relatively poor C_1 donor in the O-methylation of lignin and the N-methylation of nicotine when compared with methionine.

Although information about the actual chemical transformations taking place in the cell wall is meager, these studies give some indication of processes at the cellular level in youthful tissues. During maturation and development, the biochemical changes which take place in the wall may show a variety of patterns. The maturation and ripening processes in fruits are, for example, closely associated with pectin chemistry.

In the peach, the parenchyma cells of the mesocarp increase in diameter to a maximum whereas wall thickness passes though a maximum and declines. The esterification of pectins also exhibits this kind of behavior. Thus, in the variety "Nectar", we find:

Stage	Circumference (in.)	Cell diameter (μ)	Wall thickness (μ)	Ester (%)
Immature	6.5	91	1.1	81
Green	7.25	130	1.3	89
Ripening	8.0–8.5	123–146	2.1–1.7	93–100
Hard ripe	8.5	142	1.8	96
Firm ripe	8.5	141	1.3	84
Soft ripe	8.5	148	1.1	58

Additional evidence also suggests that little esterified pectin is present in the meristematic tissues of very young peaches until the cessation of mitosis. A high and relatively constant level of esterified pectin is present during cell enlargement and is presumably associated with increasing methylating capacity.

During the ripening of pears it has been observed that soluble glucosan, galactan and araban increase in the pectic fraction while insoluble glucosan, galactan and araban decreased in the hemicellulose fraction. These changes suggest that a dynamic relationship exists between wall and cytoplasm during the ripening process.

One of the high points in cellular differentiation and maturation is the process of lignification. The various biochemical theories of lignification fall conveniently into three categories: (a) lignin arises in the cell wall by direct transformation of other wall components; (b) lignin arises from precursors which diffuse centripetally from their point of origin in the cambium and become incorporated into the walls of xylem and phloem; and (c) lignin arises from cytoplasmic precursors formed in differentiating cell and is subsequently incorporated in the wall.

The first hypothesis is of historic interest; direct aromatization of formed polysaccharide can be rejected without reservation. Of the remaining hypotheses, the second is the more widespread and the popular one at present, and is supported by a great deal of experimental data. This data can be represented by experiments showing that radioactive lignin arises in xylem from labeled precursors supplied to the cambium. The position of label is unchanged from precursor to residues in the lignin polymer. Further, precursor type molecules such as coniferin and syringin do indeed occur in the cambial zone in many species. On the other hand, the pattern of lignification shows several features which are more consistent with the relatively neglected hypothesis of cytoplasmic (or intracellular) origin. By ultraviolet microphotometry it is evident that lignin deposition begins at primary wall corners thence spreading through the intercellular layer and inward. Further, lignified cells or cell aggregates may be completely surrounded by lignin-free tissues. Thus, morphological and chemical observations may be adduced as evidence in support of either hypothesis. It has been proposed, therefore, that a precursor may arise in the cambium and diffuse therefrom but can undergo further necessary transformations only in particular, suitable cells. This compromise provides a biochemically and physiologically reasonable way in which the two hypotheses can be merged to account for diffusion gradients and localization patterns alike.

The processes involved in growth call for a complex array of chemical transformations, as we have seen. In elongation growth, there are also profound architectural changes in the cellulosic frame. The walls of meristematic cells characteristically possess a tubular texture which is even retained in the thin primary walls of thickened cells such as bast and cotton fiber. The tubular wall contains micellar strands lying at right angles to the fiber axis of the cell, an arrangement which gives, rise to its optically negative character. When the tubular wall is stretched mechanically, it becomes optically positive. When it is stretched during growth the negative character is retained. Only a small amount of elastic deformation (8 per cent elongation in the *Avena* coleoptile) may be required to reverse the optical character of the wall. Upon release, the original negative orientation is approximately restored. Further, mechanical deformation stretches the cell at the expense of its diameter, a condition quite distinct from wall thickening in elongating cells. To account for the unique features of growing walls, it has been suggested that the tubular framework consists of a regular network held together in part at the junctions where cellulose strands meet. If in this structure the junction points are loosened, the frame can be pulled apart readily and the cell will elongate. A loosened, more open structure will result, but can be once again rendered compact by the formation and interposition of new cellulose strands (intussusception). Weakening of inter-fibrillar forces of course renders the wall plastic, but no extension can occur without the expenditure of energy. The driving force which brings about elongation is the hydrostatic pressure exerted against the wall by its contents. The maintenance of hydrostatic pressure is a function of the osmotic potential energy and directly or indirectly, energy derived from respiratory processes. The directionality of cell enlargement can be explained if it is recalled that in the tubular wall, molecular chains and therefore primary valence forces are principally arranged at right angles to the long axis of cell. In contrast, the interchain forces arranged parallel to the fiber axis must consist of far weaker hydrogen bonds. Accordingly, the wall yields preferentially in the direction of these secondary forces so that the weakly coherent strands will pull apart as the cell undergoes extension. This picture may well apply to the ideal tubular cell, but the existence of the conspicious radial enlargement which has been discussed requires a modified interpretation. The low wall polysaccharide content of young, mitotically active cells suggests a more open, less coherent wall than was realized but a few years ago. The time relations in wall synthesis and the radial–longitudinal transition support the notion that the ability to elongate depends upon formation of a more complete wall. The changeover in the enlargement process is in part a consequence of the directional weakness typical of tubular texture. Thus there must be an intermediate stage after division and prior to elongation in which the tubular character appears.

The process of surface growth may well be based upon the synthesis of microfibrils at points over the cell surface where the wall comes into intimate contact with cytoplasm. In contrast, there is an older concept of tip growth which is at variance with recent studies on radioactive glucose incorporation. It has also been suggested that the pit fields of the wall serve as centers of cellulose synthesis, but the recent concept of multinet growth seems to be of considerable value in relating wall synthesis with cell extension. By this concept, transversely oriented (optically negative) fibrils are deposited continuously as a loosely coherent layer on the inner surface of the wall during elongation. During extension growth, disorientation of the microfibrils causes them to assume a net-like form. The continuously produced transverse inner layer would tend to maintain its pattern, whereas external, older layers would exhibit progressively greater disorientation. *Avena* coleoptiles, cotton seed hairs, and other cells provide strong support for the existence of multinet growth.

Regulation of Cell Wall Formation

The synthesis of various cell wall components under the influence of 3-indoleacetic acid or other agents has been studied in several tissues. In *Avena* coleoptiles, for example, the incorporation of acetate into pectates, polyuronides and cellulose is depressed slightly, whereas the incorporation of sucrose into protopectin and cellulose is enhanced slightly by 1.8×10^{-5} M IAA. This auxin concentration has a more striking inhibitory effect on the conversion of sucrose into polyuronides, but increases both sucrose and acetate incorporation into non-cellulosic polysaccharides by about 50 per cent. Later studies in *Avena* indicate that 2.8×10^{-5} M IAA accelerates incorporation of methionine CH_3 into cell wall pectic fractions, but has no effect on acetyl content.

In wheat roots and leaves 10^{-5} M 1-naphthaleneacetic acid inhibits growth. The naphthaleneacetic acid-inhibited leaf shows increased hemicellulose and cellulose. In the inhibited root, pectin is higher, hemicellulose slightly increased, and cellulose unchanged. Similar effects can be produced in wheat by calcium deficiency. In contrast with the specific implication of auxin in pectic metabolism, the behavior of wheat has led to the view that growth regulators change the structural pattern of wall carbohydrate deposition rather than their biosynthetic reactions themselves.

A functional regulatory relationship between the pectic substances and 3-indoleacetic acid may occur in the abscission process. The physiology and chemistry of abscission are yet to be analyzed in detail, but it is clear that a breakdown of intercellular substances is involved in the loss of tissue integrity at the abscission zone. The cell wall chemistry of the abscission zone, when it is morphologically distinguishable, is sometimes destructive. In *Citrus*, for example, the cells bordering the abscission layer are highly suberized. In many species abscission may be prevented by applications of IAA or synthetic auxins and enhanced by substances or conditions which lower the internal IAA (auxin) level. Thus, directly or indirectly the hormone has an influence upon the intercellular layer.

The chemistry and biochemistry of lignins and lignification differ markedly from the transformations involved in the synthesis of wall polysaccharides. Accordingly, it is to be expected that the regulation of lignin synthesis and depositions will exhibit unique features. The control of lignin deposition by polysaccharides will be treated subsequently as a problem in surface biochemistry. Thus, the present discussion will deal primarily with control of the biosynthetic reaction. The relation between 3-indoleacetic acid (IAA) as a hormone and lignin synthesis is of paramount importance.

The effects of IAA are, however, far from simple and unidirectional. Thus, IAA induces synthesis of the enzyme peroxidase (as part of the IAA-oxidase system) and can stimulate an abnormal pattern of vascularization in pine wood—formation of highly lignified tracheids. IAA also stimulates xylem regeneration and differentiation in the *Coleus* stem. On the other hand, in experimental systems where peroxidase is abundant and not subject to increase, the effect of IAA is that of a powerful inhibitor of lignin synthesis. This inhibitory property relates to the peroxidative reaction itself and is a sensitive reflection of the newly discovered antioxidant character of IAA.

How may these apparently conflicting views be reconciled? One answer to this question depends upon a broader view of the relation between IAA, peroxidase, growth, and lignification. Thus, it has been proposed that the IAA-induced formation of peroxidase would in itself lead to increased oxidizing power. The relatively non-specific character of peroxidase would in turn lead to increased and indiscriminate oxidation of essential metabolites, enzymes, and structural elements of the protoplast. For example, catalytically important protein tyrosyl and sulfhydryl groups, free phenols, ascorbic acid and a host of cellular constituents would be oxidized to the detriment of their biological functions. Such changes accompanying lignin formation would lead altogether to cellular dysfunction and senescence, perhaps eventually to the dead, fully lignified cell. As an antioxidant and growth promoter, IAA may suppress temporarily these potentially heightened destructive processes, thus extending the period of active growth. Eventually, the peroxidizing power of the tissues must destroy the IAA itself (as part of IAA-oxidase), removing the constraint upon widespectrum oxidation. The net effect would then be one of enhanced lignification displaced in time. This concept is supported by the variety of growth promoters which as antioxidants inhibit lignin synthesis in model systems and the growth-inhibiting oxidants which hasten formation of the lignin polymer:

Antioxidants which promote growth and inhibit lignin synthesis	Oxidants which inhibit growth and enhance lignin synthesis
IAA *iso*nicotinylhydrazine mescaline skatole Reduced diphosphopyridine nucleotide (DPNH)	H_2O_2 *p*-benzoquinone 1,4-naphthpquinose adrenochrome

In addition to the foregoing substances, the more familiar biological reductants such as ascorbic acid and glutathione and dehydrogenases have been proposed as inhibitors of lignin synthesis. Metabolites which affect cellular reducing power including glucose-1-phosphate, glycerol, Krebs cycle acids and adenylic acid also inhibit the experimental formation of lignin. Finally, more generalized inhibitions have been attributed to aspartic acid, arginine, lysine, glycine, Ca (II) and Mg (II). In general then, those substances characteristic of the young growing cell with high levels of hydrogen transport activity suppress lignification.

Incomplete as our present knowledge of cell wall regulation may be, there are nevertheless the outlines of a highly functional process relating growth, differentiation, hormones and metabolism.

II. THE LYSIS OF CELL WALLS

ALTHOUGH they are somewhat removed from the intensive activities of the protoplast, cell walls are nonetheless dynamic cellular constituents, subject to the action of various hydrolytic enzymes. Enzymic attack upon cell wall substance may originate within protoplast itself, or may result from the action of enzymes released by other cells or organisms.

Functionally, cell wall lysis may be associated with several kinds of biological behavior or interactions among organisms:

(a) Resorption of formed wall structures occurs during growth and development. In insects, the production of new cuticle under the preexisting layers involves the release of chitin- and protein-hydrolyzing enzymes. In plants, sculpturing of cell walls and dissolution of end walls are an integral part of vascular differentiation: The ripening and softening of fruits involves the autolysis of intercellular substance by pectic enzymes; and in the process of abscission a weak end zone is formed, also through the dissolution of the pectic substance.

(b) Lysis of specialized walls is a part of the process of spore germination in many organisms.

(c) Invasions of cells or tissues by parasites depends in part on the lysis of host walls or other intercellular materials by the invader. Phytopathogenic bacteria and fungi, particularly "soft rot" organisms, possess a full complement of enzymes for degrading intercellular pectic substances. Animal pathogens frequently release hyaluronidase, an enzyme which hydrolyzes the important connective tissue ground substance, hyaluronic acid.

(d) Among the factors which influence the survival of individual species or strains of micro-organisms in mixed populations, the release of various hydrolases must be included as one of considerable importance. The production and release by specific organisms of wall or capsule-lysing enzymes are assumed to be independent of the presence of "competing" forms, and may sometimes originate in autolysis of individuals, yet could affect the survival of the source organism and the constitution of the associated population. Wall-lysing enzymes in multicellular organisms may serve in a defensive capacity by attacking invading pathogens in susceptible tissues. Such enzymes (lysozyme, for example) may be already present in the inter- or extra-cellular mileau or may be released by host cells which have been lysed by the parasite.

(e) Finally, wall-lysing enzymes may be of importance at a nutritional level. In herbivorous mammals, the major components of vegetable foodstuff are, of course, cell wall substances. These animals are dependent upon the microflora and fauna of the digestive tract for their cellulolytic activity. In rumen fluid bacteria play a major role in cellulose digestion, oligotrich flagellates a lesser role. In termites, as in mammals, the ability to digest cellulose depends upon intestinal micro-organisms. Again, both bacteria and flagellates are present. Land mollusks, on the other hand, produce cellulose (and chitinase) themselves, and utilize nutritionally the hydrolysates formed by these enzymes.

Many micro-organisms produce similar lysins which seem to be enzymic, but have not yet been characterized. Among these are many oxygen sensitive enzymes, which may be converted from inactive to active forms reductively. Such enzymes may account in part, at least, for anaerobic autolysis in bacteria (*Bacillus subtilis* strains, for example).

During the germination of *Bacillus* spores (e.g. *B. megaterium*, *B. cereus*), a diamino pimelic acid-alanine-glutamic acid-hexosamine peptide of *MW* ca. 10,000, other peptides and free amino acids, are released. The activated spore seems to contain a lytic enzyme (or enzymes) similar in a broad sense to lysozyme, although clearly not identical with it.

Among cell wall hydrolases in general, both chitinase and cellulase are distinctive as enzymes which attack insoluble substrates.

The residues from enzymatic attack upon celluloses become progressively more resistant to further enzymatic hydrolysis. If, however, these residues are swelled in phosphoric acid, they once again, become susceptible to cellulase. Increased resistance in residues and restoration of susceptibility correlate well with the increased crystallinity of hydrolytic residues and the disordering effect of phosphoric acid. It seems clear, therefore, that cellulase attacks amorphous regions in celluloses where interaction between functional groups in the polysaccharide and solvent or other polar molecules will be strongest.

Both cellulase and chitinase are activated by small amounts of additives. Cellulase is activated by crystalline bovine plasma albumin, β-lactoglobulin, pepsin, lysozyme, and gelatin (at 50 mg/l). Albumin, at 30 mg/l. increased cellulolysis five-fold. Similarly bovine serum albumin increases chitinases actively two-to five-fold.

The discussion, thus far, has indicated the significance and unusual properties of some wall-lysing enzymes. To conclude this consideration of cell wall breakdown we will review briefly the important representative lytic enzymes, their reactions and some organisms in which they have been found.

Cellulase:

cellulose→oligosaccharides, cellobiose, glucose (mollusks, annelids, nematodes, protozoa, bacteria, fungi, vascular plants).

Levanpolyase:

levan→oligosaccharides such as 1,6-difructosyl fructose and 1-levanbiosylfructose (bacteria).

α-Pectinglycosidase:

high ester pectin→acid-soluble uronides (fungi, vascular plants).

β-Pectinglycosidase:

pectin→mono-, di-, tri-galacturonic acid (fungi, vascular plants).

Pectinglycosidase:

high ester pectin→alcohol-soluble uronides (bacteria).

Pectin methylesterase:

pectin→pectic acid (bacteria, fungi, vascular plants).

Hyaluronidase:

hyaluronic acid→N-acetylamino-D-glucosyl-glucuronic acid, (bacteria, animal cells).

Chitinase:

chitin→N-acetylglucosamine (bacteria, fungi, mollusks, insects).

Lysozyme:

hexosamine-peptide cell wall→soluble hexosamine-peptides and hexosamine-muramic acid dimer (bacteria, mammals).

Some care must be exercised in deciding whether or not an enzyme present in animals with gastrovascular systems is in fact a product of the animal or of the micro-organisms in its gut. Thus, some uncertainty remains as to the cellulase and chitinase in mollusks and other invertebrates.

Other carbohydrates such as glucuronidases, glucosaminidases, cellobiase, etc., would also be included in an exhaustive listing. The presence of lipids in many walls may also justify the inclusion of appropriate enzymes which split the various kinds of ester linkage. Conventional proteolytic enzymes do not ordinarily attack the bacterial wall, its peptide content notwithstanding. Either the inaccessibility of requisite peptide sequences or the high proportion of D-isomers might contribute to the resistance of bacterial walls to proteolysis.

Our broad view of wall substances calls for inclusion of collagenase, an enzyme associated with the resorption of bone by osteoclastic cells, and the proteolytic enzyme(s) which degrade arthropodin in the insect cuticle.

Lignin is one of the most resistant cell wall substances, but it is slowly degraded by some wood-destroying fungi. As an aromatic polymer, its degradation must proceed through oxidative rather than hydrolytic cleavage.

III. Surface Processes in Lignin Polymer Formations

The synthesis of lignins in tissues has been studied in various ways including its overall biogenesis from carbohydrate-level constituents; the incorporation of radioactivity in the naturally formed substance following the feeding of carbon-14-labeled precursors such as tyrosine phenylalanine, etc., or by the "forced" production of lignin-like polymers from suitable pre cursors.

Each technique has contributed significantly, and uniquely, to our concepts of the physiology and biochemistry and chemistry of lignification, and each possesses limitations as well.

A recent method of particular interest has centered about forced, rapid synthesis of lignins from "Baustein" molecules in the presence of high concentrations of hydrogen peroxide. This procedure permits phenolic substances which may not be identical with the native precursors to undergo transformation into lignins. Nevertheless, a great deal has been learned by the application of such molecules in model systems which could not have been made evident by alternative methods of study. As we have already learned, the rapid synthesis of lignin from, say, eugenol, is effected by the cell wall itself rather than soluble or particulate components of the protoplast. How shall we interpret such a finding? One explanation which might be put forth is that the cell wall contains a necessary biocatalyst bound, perhaps, to its very framework. Although appreciable peroxidase resides in the washed cell wall, most of this enzyme is located in the soluble cytoplasmic fraction. We must conclude that this enzyme, which may be essential to polymer formation in our model, is but one of the requisite parts of the signifying system. We may formulate this in the following manner:

$$\text{eugenol} \xrightarrow[\text{H}_2\text{O}_2]{\text{peroxidase}} \text{intermediate} \xrightarrow[\text{reaction}]{\text{cell wall}} \text{lignin}$$

Perhaps several steps intervene between eugenol and lignin, the formation of native precursors such as coniferaldehyde, for example, and the generation of radicals from this molecule. Such intermediary steps could be effected by peroxidase together with non-enzymic reactions. Thus eugenol would not itself be rigorously a monomer in lignin formation althoug still a precursor.

In somewhat broader terms, we may reformulate our equation as:

$$\text{precursor ("premonomer")} \xrightarrow[\text{etc.}]{\text{peroxidation}} \text{monomer} \xrightarrow{\text{peroxidation}}$$

$$\text{monomer radical} \xrightarrow[\text{reaction}]{\text{cell wall}} \text{polymer.}$$

The notion that a derivative of eugenol which serves as the monomer is "activated" by being further oxidized to a radical is in keeping with the evidence for a free radical mechanism in lignin formation discussed earlier and with general concepts of free radical-induced polymerization. The above formulation suggests that the cell wall reaction is, in part, the combination of radicals to yield a polymer.

Now, it is in order to inquire into the processes underlying our so-called cell wall reaction. It is possible, for example, that the wall is the repository for a polymerizing enzyme. On the other hand, the wall may serve in some physical capacity to bring together the reactive intermediate species.

The experimental search for a distinction between these possible alternatives led into unexpected avenues of cellular biology and made possible in turn an appreciation of the great controlling forces for chemical reactions which reside in organized macromolecules themselves.

If lignin synthesis is limited by an enzymic cell wall reaction the overall process should be sensitive to typical protein-inactivating agents such as heat or urea. Unfortunately, such treatments also destroy the catalytic activity of cell wall peroxidase, which is recognized as an essential participant in the lignin polymer formation. In contrast to a direct test of the proposition which implicates a specific enzyme, the possible physical participation of the cell wall can be tested more directly, by recourse to a suitable model for cell wall substance. Logically, attention was first focused upon the major frame substance, cellulose, which is also known to be an active adsorbent.

The general scheme for the conversion of precursors to polymer calls for the participation of peroxidase. Accordingly, the simplest cell wall model system to be devised included eugenol, peroxidase, hydrogen peroxide and cellulose which was supplied as filter paper.

In practice, filter paper was simply immersed in an aqueous solution of the reactants for a suitable time and then tested for the presence of lignin. This kind of experiment revealed that traces of lignin-like materials were indeed deposited in the paper. Although the yield was disappointingly small the general scheme received sufficient support to call for further study.

It will be recalled that a small but significant proportion of the tissue peroxidase was retained in the wall after destruction of the cells. Thus, the active cell wall preparations studied actually contained peroxidase when they were placed in eugenol-peroxide solutions. This condition was not met by the first filter paper model, but was mimicked by permeating pure filter papers with crystalline horseradish peroxidase and allowing them to dry in the cold. When such treated papers were immersed in eugenol-peroxide solutions, that is, handled as if they were washed cell wall isolates, yields of lignin ranging from 12–25 mg per gram of paper were obtained. The soluble product isolated from the filter paper yielded on analysis, 63–64 per cent C, 6–7 per cent H, and 15–16 per cent OCH_3. From such analytical values, ultraviolet absorption spectra, color reactions, and solubility, it was ascertained that a material very much like gymnosperm lignin had been formed. Other celluloses, including cotton and milkweed fibers (seed hairs of *Asclepias* sp.) are similar in behavior to filter paper.

Such experiments indicate the value of highly simplified model systems in the study of lignification, and perhaps, other processes as well. They follow the reaction schemes which have been proposed and support the notion that lignification is completed by a physical interaction at the cell wall.

Although cellulose was a reasonable first choice for this study, we know it to be only one among several wall components, including protein, pectic substances, and other polysaccharide derivatives. It is of some importance to know how specific the requirement for cellulose is in model lignification systems. In the absence of cellulose, the otherwise complete system yields simpler products but no lignin polymers. Soluble proteins including gelatin and peroxidase itself, and fibrous proteins such as collagen, hair and fibrin all help to promote the formation of lignin.

Simple carbohydrates, glucose, sucrose, raffinose, similarly show no activity. In early experimental studies, it was decided to test a substance closely related to cellulose, i.e. chitin. Chitin is the polymer of 2-acetylamino-β-D-glucopyranoside hence is an analog of cellulose, with similar molecular and crystallographic properties. Native chitin isolated from the egg case of the mollusk *Busycon* showed only trace activity until deacetylated by alkali treatment, giving thereafter three-fold more lignin than cellulose itself. A consideration of the disaggregating effect of alkali upon chitin led to tests with soluble polysaccharides, starch and methyl cellulose, both highly active in lignin synthesis. These polysaccharides were from 50- to 250-fold more active than cellulose. When it was tested subsequently, pectic acid exceeded cellulose in activity by a factor of approximately 100.

The activities of the several effective substances were compared with cell wall preparations from pea root tissue. Cellulose exhibits one-twentieth and chitin one-seventh the activity measured in cell wall systems, whereas pectic acid and methyl cellulose were found to exceed cell wall preparations 2–10-fold.

Hence, the net activity of the cell wall may involve a comparatively small contribution from cellulose itself together with large contributions from polysaccharides of lower molecular weight such as pectic acid.

The relative contributions of cellulose and pectic substances to lignifying "capacity" of the cell wall is well illustrated by a different sort of experimental test. Returning to more biological models, the vascular strands of celery (*Apium graveolens*) have been employed as efficient lignifying tissues. When these strands are pretreated with the enzyme cellulase prior to use, the yield of lignin under standard conditions was reduced by one-third. When the strands are treated with the enzyme pectinase their yield of lignin was lowered by more than two-thirds. The pectic acid model also agrees with other histochemical facts. Thus, the middle lamella and parts of the primary wall which are rich in pectic substances lose their affinity for the specific stain ruthenium red as lignification proceeds. Further, during differentiation of wood fibers, the same ruthenium red-staining regions are the first to show the presence of ultraviolet absorbing substances and lignin color reactions.

Having established that the sequence of reactions in the lignification process involves the preparatory action of peroxidase followed by a polysaccharide–directed polymerization, we may now inquire into the nature of this polysaccharide–dependent cell wall reaction. Chitin which formed only minute quantities of polymer in its native state, was markedly increased in activity after removal of acetyl groups. Similarly, partial acetylation of cellulose reduces its activity substantially; when only one-fourth of the cellulosic hydroxyl groups are thus blocked, more than half of the lignin-forming capacity is removed. Thus, acetylation introduces a blocking group which abolishes lignifying capacity out of proportion to the percentage of functional groups thus obscured. Clearly, an uninterrupted polysaccharide surface is of importance in the reaction.

The highly active methyl celluloses when mixed with eugenol and some other phenols form gel-like precipitates from which the phenol may be recovered by simple solvent treatment. Thus, complex formation occurs between representative reactants.

These findings, together with the established adsorbent properties of polysaccharides and the known interactions of sugar and phenolic hydroxyl groups all point to the importance of hydrogen bonding in the precursor- or monomer-polysaccharide interaction.

The experimental behavior of other polymeric substances such as asbestos minerals and Dowex-50 provide further evidence for a physical interaction of the general sort discussed. The asbestos silicate chain resembles a distorted cellulose chain in which oxygen atoms can serve as H-bond acceptors. Dowex combines in a single high molecular weight substance the polymerization–directing capacity of polysaccharides with the catalytic activity of a peroxidase-mimicking polymer.

The experimental picture which may, therefore, be constructed places cell walls in a strategic position in the processes of lignification. Their role is somewhat unique insofar as they favor the process of polymerization on their surfaces while the mileau of a homogeneous reaction solution favors dimerization instead. They clearly lack the high order of specificity commonly associated with the genic "template", but in consideration of their limited directive properties, seem to act as a "matrix", and have been so named.

CHAPTER 3

COMPARATIVE CHEMISTRY OF INTER-CELLULAR SUBSTANCES AND WALLS

I. Phylogenetic Aspects of Lignins and Lignin Synthesis

Among the secondary substances of the plant kingdom, lignin is one of the few which have been assigned a definite role in evolution. The mechanical integrity of the higher plant cell has been associated with its lignification. The deposition of lignin upon and within the polysaccharide framework may be expected to increase considerably its resistance to compression forces, allowing development of more massive structures. Thus, the capacity for lignification has been implicated in the rise of the upright vascular land plant.

A close relation between the compression force generated by the mass of the plant and its lignin content has been established, even for a small woody stem (*Euphorbia*, Fig. 1). In contrast, a hollow-stemmed *Equisetum* shows no correlation between silica or combined silica–lignin content along an axis of comparable length. In the hollow structure, the pressure exerted by the plant upon itself is approximately 700 dyne-cm^{-1}, one-tenth of the figure obtained for the solid woody stem.

Such considerations, emphasizing as they do the evolutionary significance of lignins, call for further knowledge of the phylogenetic features. These features may be studied in two ways: (a) by chemical measurements on fossil plant materials; and (b) by chemical and biochemical measurements upon living members of representative taxa.

We shall consider first the second of these approaches, in which experimental biosynthesis of lignin plays an important part.

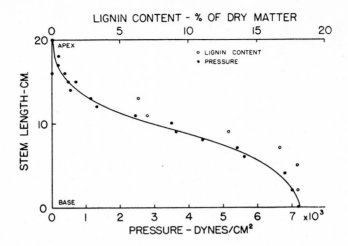

FIG. 1.   Pressure–lignification diagram for *Euphorbia*. (From Siegel, S.M., The biochemistry of lignin formation. *Physiol. Plautarum* **8**, 20, 1955.)

Tissues from seven angiosperms and fifteen other plants have been used in the comparative study of lignification (Table 1).

Among these species, thirteen contain lignin, although several have only trace quantities and several show typical lignin color reactions in atypical tissues. Thus lignin is conventionally located in vascular tissues, but occurs only in the rhizophore in *Selaginella*, in the peristome teeth of the *Polytrichum* (sporophyte) capsule, and in the sporophore of *Polyporus*. Among the vascular plants, the aquatic form *Elodea* is conspicuous for its virtual lack of lignin.

The oxidases of importance in the lignification process include both the hemoprotein peroxidase (Px) and the copper-containing phenolases (Ph). Their most important physiological distinction lies perhaps in the electron acceptor (oxidant) involved—H_2O_2 or O_2. Some of the copper oxidases, potato tyrosinase for example, cannot convert eugenol into lignins.

All of the chlorophyllous forms contain peroxidase alone (9 spp.) or both types of oxidizing enzymes (6 spp.). Two fungi contain traces of peroxidase; five fungi contain phenolases only.

All of the angiosperms studied (save *Elodea*) fall into a common group with respect to their dioxane-soluble lignin. The lignins of this fraction give typical lignin color tests and, to the extent

TABLE 1.   CLASSIFICATION AND FEATURES OF TEST SPECIES

Phylum	Sub-phylum or class	Species	Lignin	Oxidases
Tracheophyta	Pteropsida	*Apium graveolens*	+	Px+Ph
		Cucurbita pepo	+	Px+Ph
		Solanum tuberosum	+	Px+Ph
		Mimosa pudica	+	Px
		Phaseolus vulgaris	+	Px
		Pisum sativum	+	Px
		Elodea densa	trace	Px
	Psilopsida	*Psilotum triquetrum*	+	Px+Ph
	Sphenopsida	*Equisetum arvense*	+	Px+Ph
	Lycopsida	*Selaginella rupestris*	+	Px
Bryophyta	Musci	*Bryum argentum*	—	Px+Ph
		Polytrichum sp.	trace	Px (trace)
		Rhodobryum sp.	—	Px (trace)
	Hepaticae	*Marchantia polymorpha*	—	Px
		Jungermannia sp.	trace	Px
Eumycophyta	Phycomy- cetae	*Achyla ambisexualis*	—	Px (trace)
		Phycomyces sp.	—	Px (trace)
	Basidio- mycetae	*Agaricus campestris*	—	Ph
		Armillaria sp.	—	Ph
		Boletus sp.	—	Ph
		Polyporus sulfureus	trace	Ph
		Russula sp.	—	Ph

that comparisons have been made, possess similar carbon, hydrogen, and methoxyl contents.   They are quite similar spectrophoto-metrically (Table 2).

Among the other species, *Polyporus* sporophore lignin most closely resembles that of the angiosperms, however the dioxane extracts of *Psilotum*, *Equisetum*, and *Selaginella* are not too dis-similar in ultraviolet spectrum to angiosperm extracts. Only the apparently specialized lignins of *Polytrichum* and *Jungermannia* differ markedly in their absorption characteristics.

TABLE 2. CHARACTERISTICS OF ENDOGENOUS LIGNIN IN TEST SPECIES

Species	Location of lignin	Ultraviolet spectra of dioxane extracts		
		$\lambda_{max.}$	$\lambda_{min.}$	Density max. / Density min.
Apium	Vascular ⎱			
Curcurbita	Vascular ⎮			
Solanum	Vascular ⎰	275–280	265–270	1.1–1.2
Mimosa	Vascular ⎱			
Phaseolus	Vascular ⎮			
Pisum	Vascular ⎰			
Elodea	—	—	—	—
Psilotum	Vascular	275	250	1.9
Equisetum	Vascular	280	260	1.2
Selaginella	Rhizophore	265	250	1.1
Bryum	—	—	—	---
Polytrichum	Peristome teeth	285	280	1.03
Rhodobryum	—	—	—	—
Marchantia	—	—	—	—
Jungermannia	Rhizoids	320	300	1.25
Achyla	—	—	—	—
Phycomyces	—			
Agaricus	—	—	—	—
Armillaria	—	—	—	—
Boletus	—	—	—	—
Polyporus	Sporophore	270	265	1.2
Russula	—	—	—	—

Under experimental conditions the angiosperms convert eugenol into lignin only in the presence of peroxide (Table 3) whether or not they also contain phenolase. Complete peroxide dependency is also true for *Psilotum, Selaginella, Rhodobryum* and *Jungermannia*, relative dependency was observed in *Equisetum, Polytrichum* and *Marchantia*. Only *Bryum* and *Boletus* formed lignin well without peroxide, which, interestingly, actually inhibited the latter. The ultraviolet absorption and other features such as color reactions and chemical constitution again place the angiosperms in a common group, but the product resembles gymnosperm rather than native angiosperm lignin in all respects.

TABLE 3. EXPERIMENTAL LIGNIN FORMED FROM EUGENOL IN TEST SPECIES

Species	Lignin formed		Ultraviolet spectrum of dioxane extract		
	Without H_2O_2	With H_2O_2	$\lambda_{max.}$	$\lambda_{min.}$	Density max, / Density min.
Apium	—	+			
Cucurbita	—	+			
Solanum	—	+			
Mimosa	—	+	280–285	270–272	1.1
Phaseolus	—	+			
Pisum	—	+			
Elodea	—	+			
Psilotum	—	+	275	250	1.9
Equisetum	trace	+	282	265	1.3
Selaginella	—	+	281	255	1.3
Bryum	+	+	280	260	1.4
Polytrichum	trace	+	287	275	1.4
Rhodobryum	—	+	280	270	1.2
Marchantia	trace	+	290	260	4.2
Jungermannia	—	+	280	270	1.1
Achyla	—	—	—	—	—
Phycomyces	—	—	—	—	—
Agaricus	—	—	—	—	—
Armillaria	—	—	—	—	—
Boletus	+	—	285	270	1.1
Polyporus	—	—	—	—	—
Russula	—	—	—	—	—

We know, however, that angiosperm lignins are more highly methylated than their gymnosperm counterparts, a condition which is explained by the presence of the syringyl—or pyrogallol dimethyl ether—group in the former. Hence, the synthetic lignin reflects in its properties the precursor rather than the tissue in which it was formed.

That tissue-specific factors tend to change the product formed from a single precursor, however, is shown by the spectral prop-

erties of the eugenol lignin formed in *Psilotum* and *Marchantia*.
The synthetic lignins of *Polytrichum*, *Rhodobryum*, *Jungermannia*
and even the basidiomycete *Boletus* are virtually identical spectro-
scopically with the substances formed in angiosperm tissues. *Equisetum*, *Selaginella* and *Bryum* give similar products. Only the
first-mentioned species differ markedly, although one must conclude
that commonness of precursor by no means limits the product
formed and may allow for considerable latitude in the specific
nature of that product.

The existence of such variation as a reflection of matrix effects
may be seen in a comparison of dioxane-soluble eugenol lignin
formed on chitin (λ max 270 mμ), cellulose (λ max 280 mμ) and
the mineral amphibole (λ max 315 mμ).

A comparison between native and eugenol lignins shows us
that they are spectrophotometrically identical in *Psilotum*; somewhat different (as discussed) in the angiosperms and *Equisetum*;
and moderately to greatly different in other cases.

Of the twenty-two species studied, nine contained conspicuous
quantities of lignin and four only trace quantities. Experimentally,
fifteen species could form appreciable lignins from eugenol when
supplied peroxide and one additional species did so in the absence
of peroxide. Of the seven fungi, only one converted eugenol to
lignin. *Polyporus*, which actually contained lignin in the sporophore, could not synthesize it under any experimental conditions.
Such paradoxical behavior may be accounted for by the additional
facts that the organism was actually found thriving on a fallen
oak, hence may contain derivatives of host lignin either intact or
repolymerized from solubilized fragments which had diffused into
the mycelium.

What can be learned about lignin phylogeny, then, from such
experimental studies? First, we can recognize a most important
point, namely that the capacity to form lignified cell walls exists
in species which lack lignin or contain limited amounts in highly
specialized tissues. As a corollary, we may note that this potential
goes beyond the vascular plants, although it may not exist commonly
in the fungi.

Second, among the angiosperms, lignification is a constant feature, save in the case of the aquatic *Elodea* which produces appreciable lignin only when it is supplied with precursor. This may be considered a secondary characteristic resulting from a genetic loss in the antecedents of such plants.

Such a viewpoint is supported elsewhere in the Pteridophyta: These plants characteristically have well lignified vascular tissues, however exceptions include the water ferns *Salvinia* and *Ceratopteris*. The former, of course, occurs as a surface aquatic. Although *Ceratopteris* extends its axis well into the air, its succulent tissues are supported by hydrostatic forces.

Third, chemical diversity among the lignins obviously depends upon the chemical constitution (methoxyl content, for example) of the precursors available. In addition, however, tissue factors may modify the course of oxidative polymerization of a single precursor in a way that allows different products to be formed. The matrix phenomena noted above and discussed elsewhere. are offered as a possible example of tissue factors.

The presumption that precursor supply, precursor structure and/or tissue factors operate among different tissues as well as among different species is supported by instances of lignins localized to special tissues such as peristome teeth, and by observations on the variations (increases) in the methoxyl content of lignins with age.

Current concepts in aromatic biochemistry suggest that biogenetic relationships exist among the several major types of phenolic carbon skeletons:

$$\text{carbohydrate} \rightarrow \begin{array}{l} \text{pre-aromatic} \\ \text{pool} \end{array} \begin{cases} & C_6-C_3 \begin{cases} \nearrow C_6-C_3-C_6 \text{ flavonoids} \\ \rightarrow (C_6-C_3)_2 \text{ lignanes} \\ \searrow (C_6-C_3)_n \text{ lignins} \end{cases} \\ \uparrow\downarrow \; C_2 \\ \searrow C_6-C_1 \longrightarrow \text{ tannins, depsides} \end{cases}$$

In this scheme, substances at the carbohydrate level must undergo cyclization and desaturation, forming intermediates such as prephenic acid, and in turn phenols with one—or three—carbon side chains. These, in turn, may lead to several terminal products including flavonoids with an additional phenolic group; the dimeric lignanes; the polymeric lignins; or various benzoic acid derivatives such as the tannins. Further, compounds bearing C_1 and C_3 side chains may be linked biogenetically through acyl group transfer ($C_1 \rightarrow C_3$) or β-oxidation ($C_3 \rightarrow C_1$).

Where, in such a scheme can we find the most likely place for the genetic events leading to a supply of lignin precursor? In some forms tyrosine can serve as a precursor, in others it does not seem to be suitable. If we make the reasonable assumption that all of the organisms considered necessarily produce tyrosine as a part of their protein anabolism, it follows that they must have some capacity for making the $C_6 - C_3$ precursors of lignins. *Elodea*, for example contains tyrosine and flavonoids (as anthocyanins), but little or no lignin.

In this case, at least and likely in others as well, we may be confronted with an example of competing metabolic pathways. Thus, the potential precursors are not realized as they are transformed into other products more effectively (rapidly) than they can be converted to lignins.

The origins of lignin synthesis are yet obscured, but it is possible to suggest that the genetic basis for lignification resides in at least four areas:

(a) Generally, but not exclusively, in the ability to synthesize peroxidase.

(b) In quantitative factors controlling the rate of phenylpropane synthesis (relative, of course, to the rate of phenylpropane consumption).

(c) In specific factors controlling hydroxylation patterns and methyl transfer.

(d) In specific factors controlling the composition of the polysaccharide framework (hence matrix properties) of the cell wall.

Finally, lignin synthesis is clearly an aerobic process. Although the precursors could be fermentation products, peroxidases are aerobic, H_2O_2 is formed by reduction of O_2 and, of course, the phenolases utilize O_2 directly as an electron accepter. Hence, precursor synthesis may have proceeded in some forms during part of the long phase preceeding formation of an oxidizing atmosphere. Accordingly, the littoral ancestors of the upright land plant may have been "primed" for lignification before free oxygen became a significant part of their environment.

Before ending our phylogenetic considerations, the status of lignins in fossils themselves deserves brief review. The lignins of angiosperms are distinguished from others by their high methoxyl content. Analytically, the lignins of gymnosperms and other "lower" forms yield, on nitrobenzene oxidation, p-hydroxybenzaldehyde and vanillin (4-hydroxy-3-methoxy-benzaldehyde). Angiosperm lignins yield, in addition, highly methylated syringaldehyde. In lycopsids, sphenopsids, and older members of the Pteropsida, vanillin-yielding lignins may be traced at least to the Devonian. Vanillin-yielding moss remains seem to have originated in the Upper Carboniferous.

The syringyl group is predominantly associated with the Cretaceous angiosperms but, interestingly, is also found in the gnetales, dating from the Jurassic.

Betulinium wood (Pliocene-Miocene) gives intense lignin color reactions and assays about 35 per cent lignin with a methoxyl content of 11·4 per cent *Podocarpoxylon* lignite from the upper cretaceous contains 69 percent lignin containing about 7 per cent OCH_3.

Among Carboniferous forms, specimens of *Lepidodendron* and *Calamites* contain small amounts of dioxane-soluble materials giving the characteristic lignin color reactions. *Lepidodendron* extracts also have a typical (non-angiosperm) lignin ultraviolet spectrum.

Such evidence shows clearly that the lignins as chemical entities have considerable antiquity, dating back to the period associated with the origin of the land plant.

Although lignins are comparatively resistant to the conditions associated with simple burial of vegetable matter, they are nevertheless modified in fossils. Decline in methoxyl content is one such change, but more extensive modification to the aromatic nucleus also occurs.

If, for example, we examine related fossil and modern forms for extracts absorbing in the aromatic region of the ultraviolet, we find the number of fractions and their intensity to be markedly reduced (Table 4). The apparent loss of aromatic content may involve many factors, but two changes in lignins are known to be brought about by elevated temperatures (157–230°C) in the absence of oxygen. First, the characteristic ultraviolet absorption curve of lignins is altered, becoming more flattened and less distinctive, and second, purified dioxane—and alkali-soluble lignin becomes progressively less soluble with increasing severity of heat treatment. At 230°C, for example, dioxane solubility is reduced three fold in 20 min. Even at 100°C, lignin treated in air for 10 days undergoes some loss of solubility, a marked shift in absorption maximum (280 mμ→310 mμ) and loses some 80 per cent of its absorbancy per unit weight.

TABLE 4. ULTRAVIOLET ABSORBING SUBSTANCES IN EXTRACTS
FROM MODERN AND FOSSIL FORMS

Solvent	Lycopsida		Sphenopsida	
	Selaginella	*Lepidodendron*	*Equisetum*	*Calamites*
HCl (36%)	+ +	−	+ +	−
HCl–ethanol	+ + +	−	+ +	−
KOH (5%)	+ + +	−	+ + +	−
5% KOH–ethanol	+ + + +	−	+ + + +	−
Methanol	+ + +	+	+ + +	−
Ethanol	+ + +	+	+ + +	+
Dioxane	+ + +	+	+ + +	+
Chloroform	+ +	−	+ +	−

The effects of such elevated, but not extreme temperatures suggest the importance of anaerobic pyrolysis in modifying the fossil lignins.

II. NON-VASCULAR PLANTS, PROTISTA AND METAZOA

THE traditional approach to the study of cell walls views them as finished morphological entities external to the life processes of their cells of origin, or as an inert framework for the support of many protoplasts. This essentially static viewpoint has been supplanted by a broader picture of extracellular and intracellular substances as participants in and reflections of cellular processes. We may treat the cell wall as a member of a great class of substances which are formed by a physiologically unique (and important) part of the protoplast—its surface. Nor need we require for our purposes that a cell elaborate particular products homogeneously distributed at all of its surfaces. Certainly this situation does not fit the epidermal cells of the green plant which deposit specialized waxes and other cuticular materials on one face only.

If the functions of cells are to be understood, full account must be taken of the many extracellular and intercellular substances produced by organisms, with the restriction, for present purposes, that these products retain a spatial relationship to the protoplast surface from which they originated.

In the search for broad principles upon which to base an understanding of the cell wall, we must examine the array of substances that various organisms produce at their surfaces without taxonomic restriction.

In pursuit of our objectives, we will have the opportunity to examine a sample ranging from the enterobacteria to Mammalian bone.

Bacteria and Fungi

The bacterial protoplast is characteristically enclosed within a rigid wall of complex constitution.

The isolation of bacterial wall substance has only recently been accomplished by combined procedures including mechanical disruption, washing, centrifugation, and enzyme treatment (trypsin, ribonuclease, pepsin). The purity of wall isolates may be established by electron microscopy, which reveals cytoplasm-free preparations resembling collapsed balloons.

Analysis shows that such preparations commonly contain lipids, ester phosphate, hexoses, hexosamines, and amino acids.

More specifically, major cell wall constituents include the following:

Amino acids

alanine	aspartic acid
glutamic acid	glycine
diamino pimelic acid	serine
lysine	

Hexoses and amino sugars

glucose	N-acetyl glucosamine
glucosamine	muramic acid (3-O-D-carboxyethyl hexoseamine)
galactose	chondrosamine
rhammose	
mannose	
arabinose	
galactosamine	

Interestingly, some bacterial walls are relatively rich in the D-isomers, of alanine and glutamic acid. In *Streptococcus faecalis* for example 25 per cent of the alanine and 85 per cent of the glutamic acid exist in the D-form. In *Lactobacillus*, over half of the alanine and two-thirds of the glutamic acid are D-isomers. Although half of the alanine from the wall of *Staphylococcus aureus* occurs in the D-form, the glutamic acid is almost entirely of the D-form. In general, D-alanine is most widespread among various bacterial walls, D-lysine is also known to occur, but diaminopimelic acid occurs most frequently as the *meso*-form.

These and other constituents are in turn organized into polysaccharides, polymeric hexosamines, peptides, peptide–polysaccharide complexes, proteins (including wall antigens), lipoproteins, and glycoproteins.

Among the peptides found in various bacteria, the combinations alanine–glutamicacid–diaminopimelic acid or alanine–glutamic acid–lysine recur, sometimes with aspartic acid, glycine or serine in addition. Cyclic peptides containing only lysine and aspartic acid have also been demonstrated, but may be artifacts of acid hydrolysis. Thus far, the presence of either of diaminopimelic acid or lysine in a particular instance precludes the presence of the other.

The fundamental cell wall components are further organized into systems of molecules and macromolecules. The basal structure of the wall in Gram-positive forms commonly consists of: (1) a hexosamine polymer derived from N-acetylglucosamine, muramic acid and galactosamine; (2) the peptides, alanine–glutamic acid––lysine or alanine–glutamic acid–diaminopimelic acid; (3) polysaccharides; (4) phosphorylated compounds; (5) antigenic proteins. The hexosamine polymer chains provide an insoluble ground substance to which the peptides and other components are attached in an as yet uncertain manner.

Gram-negative forms such as *Escherichia coli* contain similar basal structures. The rigid basal structure is, however, "masked" by a lipoprotein layer. Upon removal of the lipid material, the characteristic Gram-positive wall can be demonstrated.

Glucosans or aminoglucosans are evidently commonplace components of the bacterial cell wall (although *Corynebacterium haemolyticum* has no glucose in its wall) whereas the occurrence of α-cellulose as a major wall substance is unusual. *Acetobacter xylinum* when incubated with hexoses, D-fructose and D-glucose especially, elaborates a wall substance which is indistinguishable from natural cotton cellulose. Acetylation yields cellobiose octa--acetate; methylation and hydrolysis yields 2,3,6-trimethyl-D-glucose; and x-ray analysis yields the usual diffraction pattern of cellulose.

Acetobacter cellulose consists of fibers 100 Å in thickness, 5000 Å in width, and 40μ in length. These fibers contain a weave of fibrils 200 Å in width.

Beyond the rigid cell wall, some bacteria also produce mucilaginous encapsulating polysaccharides. The ultimate building units obtained upon hydrolysis of capsular material include D-glucose, D-galactose, L-rhamnose, pentoses, glucuronic acid, galacturonic acid, hexosamines, and N-acetyl hexosamines. These substances occur in varying proportions and combinations which together with unusual monosaccharides serve to characterize individual species or strains. Thus the capsule of *Shigella dysenteriae* contains L-rhamnose, a comparatively uncommon sugar, together with D-galactose and hexosamine. In contrast, *Corynebac-*

terium diphtheriae capsules which also yield galactose and hexose-amine, contain pentose instead of rhamnose. In *Diplococcus pneumoniae*, differences among the capsules of various cultural types may also be profound. *Pneumococcus* capsule Type I contains 28 per cent D-galacturonic acid, whereas type II is 70 per cent D-glucose. Types III and IV both consist of D-glucose and D-glucuronic acid, but in the type III capsule they occur exclusively as cellobiuronic acid units in which the glucose is joined to the glucuronic acid by a 1,3-linkage, and the glucuronic acid to the next glucose by a 1,4-linkage. In type VIII, the polyscacharide chain yields D-glucose units in addition to D-glucose-containing cellobiuronic acid units.

The actinomycetes, which are a somewhat controversial group taxonomically, have cell walls consisting of hexose, hexosamine, and amino acid units (including aminosuccinoyl-lysine), the latter in only small amounts. The aerial, but not the vegetative, mycelium has an outer layer of lipid. Grossly, therefore, such walls resemble those found in Gram-positive bacteria. This similarity in wall building units has led to arguments relating the actinomycetes to the bacteria rather than the fungi. The many variations which exist among the bacteria proper and the existence of hexose–hexosamine–protein wall systems in some fungi, suggest that the comparative analysis of cell wall components be applied with circumspection in taxonamic classification.

The cell wall chemistry of the fungi has long been a subject of controversy. The presence of chitin, for example, has been claimed and questioned for many years. It has also been suggested that cell wall components vary along broad taxonomic lines. The phycomycetes contain cellulose mainly; the ascomycetes and basidiomycetes, chitin.

Perhaps much of the uncertainty and controversy associated with cell wall constitution lies in the lack of rigorous analytical procedures and specific microchemical tests. A few examples will illustrate the utility of comparatively recent analytical methods in providing a more definitive picture of the cell wall in the fungi.

Uncertainties as to the presence of chitin in the fungal wall have been settled in some forms at least by X-ray diffraction studies.

Although the chitin of intact walls commonly gives a somewhat modified diffraction pattern, simple preparative procedures have been used quite successfully to demonstrate its presence. The sporangiophores of *Phycomyces* (phycomycete) and *Aspergillus* (ascomycete) have been shown to contain a typical α-chitin. Crystallographically, the unit cell of α-chitin in fungi, as in animals is rhombic, $9.4 \times 10.46 \times 19.25$ Å in dimensions, and contains 8 N-acetylglucosamine residues. The sugar residues are present as the disaccharide chitobiose, the analog of cellobiose in the cellulose molecule. The four chitin chains they represent are further arranged in two antiparallel sets, precisely the situation in cellulose. The fiber period corresponds to two glucopyranose lengths, as in the case of cellulose. Chitinous walls exhibit the lammelar and fibrillar organization typical of cellulosic walls. Physical analysis (electron microscopy, polarized light) of the *Phycomyces* sporangiophore shows that it consists of an outer primary wall of tubular texture, a thickened secondary wall of fibous texture and a very thin inner core of steeply spiraled texture. Thus, both spiral and stratiform structural plans are present. The *Aspergillus* sporangiophore exhibits spiral texture. Although supporting chemical data have been applied in such studies the burden of proof has been served by the physical methods of analysis.

The constitution of the yeast cell wall has been elucidated mainly by chromatographic and other chemical procedures supported by physical methods such as differential centrifugation and electron microscopy. These combined methods show that the yeast cell wall which makes up about 15 per cent of dry cell contains:

lipid   hexosamine
glucose   amino acids
mannose

Glucose is present in part as a highly branched insoluble polymer whose molecular weight is about 6,000,000. The interbranch distance corresponds to nine glucopyranose units joined in 1,3-linkages; the branch points are joined by 1,2,linkages. This material in association with protein makes up about 30 per cent of the cell wall. The rest of the glucose appears in two soluble components, both containing mannose and glucose in the ratio of 2:1. The hexosamine corresponds to a chitin content of about 3 per cent. Lipid content totals about 8 per cent and is principally in a bound form; protein content totals about 7 per cent distributed among the polysaccharides as discussed. The protein component is of special interest, as it contains an exceptionally high (2 per cent) sulfur content. The sulfur in this pseudokeratinous protein which we have seen is intimately associated with polysaccharides, is critically involved in cell division in yeasts, illustrating the identity which may be attained by architectural and functional units in the operation of life processes.

It is apparent that a phycomycete can contain chitin whereas an ascomycete can possess great amounts of hexoses and little chitin. The cell walls of other phycomycetes, *Rhizopus* and *Mucor*, contain only glucosamine units, whereas the cell wall polymers of the Aspergillacese (ascomycetes) contain galactose, glucose, and mannose in addition to glucosamine.

The class Myxomycetae (phylum Myxothallophyta), also recognized as the protozoan order Mycetozoa (class Sarcodina), consists of organisms which exist for the greater part of their life cycle in naked protoplasmic form. Formation of cellulosic or calcified cellulosic walls is associated with sporulation and with resting or sclerotial stages, which form under unfavorable environmental conditions.

Algae and Selected Protozoa

The biochemical and morphological diversity known in the algae applies to their cell wall chemistry, however, save for the Chrysophyta (Chrysomonads, diatoms) cellulose is a typical wall component in all phyla (Table 5). In the Chlorophyta, cellulose may be accompanied by polysaccharides containing galactose and glucose (*Spirogyra*); by polyuronic acids such as pectin (typical of the Chlorophyta in general); or by chitin (Cladophorales, Oedogoniales). On the other hand, volvocalian forms such as *Haematococcus* and *Platymonas* and chlorococcalian forms such as *Ankistrodesmis*, *Chlorella* and *Scenedesmus* contain little or no cellulose. The *Platymonas* wall contains galactose and uronic acid units, and traces of arabinose. Some *Chlorella* species contain a hemicellulose composed of glucose.

TABLE 5.   A COMPOSITE SUMMARY OF ALGAL WALL SUBSTANCES

	Cellulose	Non-cellulosic polysaccharide	Polyuronic acids	Polyacetyl glucoseamine	Silica
Chlorophyta	typical	present	typical	uncommon	absent
Pyrrophyta	typical	absent	present	absent	absent
Chrysophyta	uncommon	—	typical	uncertain	present
Phaeophyta	typical	typical	typical	absent	absent
Rhodophyta	typical	typical	present	absent	absent
Cyanophyta	typical	absent	typical	uncertain	absent

Some chlorophycean cell walls consist of a single cellulosic layer (*Protococcus*) but a multilayered condition is more common (Table 6). The walls of *Oedogonium* and *Cladophora* consist of an inner cellulosic layer, a pectic middle layer and a chitinous outer layer; *Mesotaenium* and *Chlamydomonas* contain only inner cellulosic and outer pectic regions; *Oedogonium* contains an inner cellulose layer, a central layer of cellulose (thin) and pectin (thick), and an outer pectic layer.

The organization of cellulose in the cell wall of *Valonia* (Siphonocladiales) has been examined by x-ray and electron microscopic procedures. The wall contains several dozen lamellae, each in turn an array of parallel fibers. In successive lamellae, the fibre orientation alternates so that an extensively crossed structure results. The cellulose fibrils are about 300 Å in diameter and extremely long (10μ or more). In *Cladophora* (Cladophorales) the chains in successive lamellae make an angle of 83° with one another. The alternation further consists of steep and flat spirals.

In the Pyrrophyta, the cellulosic wall is commonplace; some dinoflagellates (Dinophyceae), have non-cellulosic walls; most motile forms have heavy homogeneous walls composed of interlocking cellulose plates; non-motile dinophyceans also possess pectin sheaths. The Chrysophyta contain three classes and a dozen

orders, but its most important forms are probably the Bacillario-
phyceae, the diatoms. These organisms ordinarily contain pectin
and silica but rarely cellulose. The wall consists of overlapping
valves that fit as do the halves of a petri dish. The silicified layer is
laid down as an areolated or striated sheet, with patterns characteristic of the genus and species.

TABLE 6

Organism and group	Wall layer		
	1	2	3
Chlorophyta			
Zygnematales			
Mesotaenium	cellulose	pectin	
Desmidium	cellulose	cellulose–pectin	pectin
Oedogoniales			
Oedogonium	cellulose	pectin	chitin
Cladophorales			
Cladophora	cellulose	pectin	chitin
Ulotrichales			
Protococcus	cellulose		
Volvocales			
Chlamydomonas	cellulose	pectin	
Chrysophyta			
Bacillariophycae			
Navicula	pectin	silica	
Phaeophyta			
Laminariales			
Laminaria	cellulose	Ca alginate polysaccharide–sulfuric acid	
Cyanophyta			
Myxophyceae			
Gloetheae	cellulose	pectin	

In the Phaeophyta, the cell wall commonly consists of a firm
inner cellulosic region and a gelatinous outer layer of polyuronic
acid and non-cellulosic polysaccharide. Unlike the green algae, the
phaeophycean polyuronide is calcium alginate, poly-1,4-α-D-galac-
turonic acid, an isomer of pectic acid. The non-cellulosic poly-

saccharide, fucoidin, is a 1,2-α-polymer containing residues of the methylpentose, L-fucose, esterified with sulfuric acid.

The rhodophycean wall contains cellulose together with gelatinous polygalactose sulfate esters. Best known of these polysaccharides is agar, which consists of D-galactose units joined via 1,3-linkages with L-galactose sulfate. The ratio of D-galactose: L-galactose sulfate lies between 9:1 and 7.5:1.

The fundamental cellulose-pectin pattern is found in the Cyanophyta. The presence of chitin has been claimed, but continues to be in doubt.

Our consideration of the algale cell wall has encompassed several protozoan groups. A number of algal forms have also received zoological recognition as flagellates (class Mastigophora) of the sub-class Phytomastigina.

The systematic relationships among these forms may be summarized as follows:

Botanical system		Zoological system
Phylum	Class	Order (all sub-class Phytomastigina)
Cholorophyta	Chlorophyceae	Volvocina
Pyrrophyta	Cryptophceae	Cryptomonadina
	Dinophceae	Dinoflagellata
Chrysophyta	Chrysophyceae	Chrysomonadina

Many Protozoa are ameboid, or possess an external "wall", the pellicle, of uncertain properties and constitution. Mineralized structures which serve as shells or as externally formed skeletons have been developed among the ameboid forms of the Sarcodina in the orders Foraminifera and Radiolaria.

The foraminiferan shell may be composed of nitrogenous materials or siliceous plates, but is commonly calcareous in nature, and often similar in form to the molluscan shell. As they grow, the radiolarians commonly form intricate spicular structures at their surfaces. These structures may exist as a lattice work at the surface, or may consist of several concentric lattice-spheres, the innermost being "buried" within the cytoplasm hence serving as an internal skeleton. The acantharian spicular system differs from the common pattern in consisting of strontium sulfate instead of silica.

Metazoa

The cells of multicellular animals are not characteristically "walled off" on all surfaces from neighboring cells by their secretions. We often find extracellular materials which have been elaborated by the exposed surfaces of cells in the external layer of the metazoan body. Such materials commonly serve as exoskeletons. Among coelenterates, the corals (class Anthozoa) are often identified with skeletal products. In many forms, the skeleton is an ectodermal product consisting either of calcium carbonate or horny proteinaceous substances.

The exoskeleton reaches high point among the arthropods. The cuticle is a complex and highly structured system of macromolecules elaborated by the epidermal cells.

The major building materials of the cuticle are fibrous proteins chitin, phenols, and, in some cases, calcium carbonate. Arthropods characteristically produce the α-chitin which has been described in the fungi.

Other forms—annelids and mollusks—produce β-chitin, which is a crystallographic modification on of the α-form.

In insect metamorphosis, cuticular development begins with the secretion of a lipoprotein layer, cuticulin, by the epidermal cells. Following this, lamellae consisting of approximately equal parts of chitin and the protein arthropodin are laid down. The old cuticle is digested by proteinases and chitinase and reabsorbed. Shortly before molting, phenols are released and incorporated in the new cuticle and a wax layer is deposited on its surface. As the new instar emerges, a final lipoprotein layer is deposited on the surface while the phenols are oxidized to quinones which cross--link cuticular proteins, forming the final dark, hard outer layers.

The chitin and protein are internally associated in the cuticle. From X-ray analysis, it may be concluded that chitin and protein lie in planes consisting of parallel rows of oriented protein and chitin chains. The protein is of the extended, β-keratin type, hence each chitobiose repeating unit, 10.3 Å in length, would coincide with each three amino acid residues in the protein.

The final process which we will consider is, like others mentioned, distinct from the traditional botanical concept of cell wall formation. Nevertheless, osteogenesis contains all the essential elements of a wall-forming process. Bone formation begins with the differentiation of mesenchymal cells into osteoblasts, proceeds through their active deposition of the intercellular substance which makes up the organic matrix of bone, and is completed with the deposition of bone mineral on the matrix. Functionally, the osteoblast situated in the calcified interstitial substance becomes the bone-maintaining osteocyte. We will not concern ourselves with the many interactions and interconversions among connective tissue cells, but only with the intercelluar substances elaborated by the bone-forming cell. Most (90–96 per cent) of the dry, lipid-free organic matter of bone consists of the protein collagen, which is deposited in the form of fibrils (0.3–0.5μ in diameter) or fibrillar aggregates. Collagens are noted for their X-ray diffraction patterns, and form a characteristic double cross-banding with a period of 640 Å. The interfibrillar ground substance consists of amorphous mucopolysaccharides such as hyaluronic acid and chondroitin sulfate, which on hydrolysis yield glucosamine or galactosamine, sulfuric acid and glucuronic acid. The amount of mucopolysaccharide in bone is small, 0.1–0.4 per cent, but it is nevertheless an important component of the organic matrix.

The third polymeric matrix component is reticulin, a fibrous glycoprotein whose polysaccharide moiety contains mannose, galactose, fucose and, perhaps, hexosamine. It is of interest in its intermediate position relative to the protein and polysaccharide constituents. The mineral composition of bone salt is approximated by hydroxyapatite:

$$3 \ Ca_3 \ (PO_4)_2 \cdot Ca(OH)_2$$

Crystallographically, hydroxyapatite is hexagonal with a unit cell 9.4 Å $\times$ 9.4 Å $\times$ 6.9 Å.

Bone crystals are submicroscopic (representative dimensions: 500 Å $\times$ 250 Å $\times$ 85 Å). Thus, they are only a few unit cells in thickness, and possess considerable area relative to mass. The crystals lie in the ground substance, forming a periodic pattern around the collagen fibers in a definite relation to the 640 Å period. Mineralization is believed to occur as a catalyzed crystallization involving nucleation centers within the matrix.

Hydroxyapatite is a mineral commonly identified with the osteogenic activities of the vertebrates. It is part of the intercellular organo-mineral complex involved in bone formation. In the cytoplasm of the ciliate, *Spirostomum ambiguum*, which has been extensively studied because of its size and ready cultivation, calcareous particles of 0.5–3μ diameter have been observed in all but the most youthful, dividing stages. By X-ray diffraction these bodies, which possess considerable fine structure, were shown to consist of hydroxyapatite. The fundamental crystallite resembles in size the primary particles in bone. It is claimed that the mineral particles are deposited on and/or within a fibrillar nucleus, revealing thus a relationship similar to that in bone.

A Note on Antigenic Cell Wall and Related Polysaccharides

It is well known that polysaccharides determine the immunological specificities of many micro-organisms, but it has been recognized only recently that the polysaccharides of many plants and animals may also react with antibodies evoked in experimental animals by suitable bacterial antigens.

Even with the limited experimental data now available it is abundantly clear that wall polysaccharides of the vascular plant are able to precipitate the antisera formed in the presence of the capsular polysaccharides of several pneumococcal types (Table 7).

In addition to the several cell wall polysaccharides noted, types II and XIV antigens evoke antisera precipitated by synthetic polyglucoses.

In addition, almost all pneumococcus antisera are precipitated by agar, carragheenins, and other polysaccharide sulfates.

Aside from its direct immunological interests, the immunochemistry of cell wall polysaccharide components offers a valuable tool in the study of their biological and molecular properties and specificities.

TABLE 7. WALL POLYSACCHARIDES WHICH PRECIPITATE
ANTIPNEUMOCOCCAL SERA

Pneumococcus type antigen	Units in capsular polysaccharide	Polysaccharide precipitants for antisera
II	L-rhamnose, D-glucose, D-glucuronic acid	hemicelluloses or corn and flax fiber, *Azotobacter*, *Klebsiella*
III	D-glucose, D-glucuronic acid	oxidized cotton, *Azotobacter*
VII	galactose, glucose, rhamnose, amino sugar	hemicellulose of corn fiber
VIII	D-galactose, D-glucose, D-glucuronic acid	oxidized cotton, *Azotobacter*
XIV	D-galactose, D-glucose, N-acetylglucosamine	hemicellulose of corn fiber, gymnosperm arabogalactans

Conclusions

Their many variations not withstanding, certain basic themes are clearly recorded in the constitution of walls and intercellular substances. The very concept of an extracellular architecture based upon organized macromolecules provides us with the most nearly universal theme, and is in turn seconded by the concept of heterogeneity in the macromolecular building materials selected.

In the recurrent cell–polysaccharide wall, cellulose, or chitin, is accompanied by polymers of hexose and/or hexuronic acid. In other cases, polysaccharide or polyhexosamine may be associated with protein. Such combinations exist in bacteria yeasts, arthropods, and vertebrates. Commonly, these hydrophilic materials are covered by or encrusted with resistant substances, such as waxes, lipids, or minerals. Mechanical adaptations involve the incorporation of substances such as lignin or mineral. In all of these instances, an ordered high polymer is laid down, infiltrated by lower molecular weight ground substances, and finally modified (and perhaps stabilized) by cross-linking agents, hydrophobic substances, or minerals.

Thus, even in this modest sample of the biological domain, we can begin to develop some picture of the way in which chemical capabilities, macromolecular properties, and surface conditions together are reflected in the substances that organisms secrete or elaborate external to their protoplast membranes.

CHAPTER 4

PROPERTIES AND USES OF CELL WALLS AND THEIR DERIVATIVES

I. Woods, Fibers, and Plastics

CELLULOSE, uronic acid fibers and plastic cell wall materials have been used by Man in many forms from early times. Twines of flax and hemp were known to Stone Age Man, and linen fabrics were prepared by the Swiss lake dwellers of the Neolithic. The arts of spinning cotton and flax date back to Egyptian and Indian civilizations of the third millennium B.C., and refined cotton fabrics were introduced into European civilizations in the fourth century B.C. Other fibers unknown to the Mediterranean peoples, hemp, jute and ramie, for example, were known to the Chinese and Indian peoples from ancient times.

The useful natural cellulosic fibers are of diverse origin and include seed hairs, bast fibers, and leaf fibers.

The seed hair products are principally cotton and kapok. Cotton, from *Gossypium* (Malvaceae) contain 85–90 per cent cellulose, 6–8 per cent water. Its fibers are 15–35 mm in length and 20–10μ in diameter. They are flattened and twisted single cells. Kapok, from the Javanese plant *Ceiba pentandra* (Bombacaceae) contains 65 per cent cellulose, 12 per cent water and 15 per cent lignin. The flossy material is valued for insulating and buoyancy properties.

Flax (*Linum usitatissimum*, Linaceae), hemp (*Cannabis sativa*, Moraceae), jute (*Corchorus capsularis* and *C. olitorius*, Tiliaceae) and ramie (*Boehmeria nivea*), are the principal sources of bast fibers. The fibers contain 65–90 per cent α-cellulose, 8–12 per cent water and various amounts of other materials. Ramie is notable for the purity of its cellulose; jute for a high lignin content. Flax cells run 6–70 mm in length and 15–20µ in diameter, but the cells overlap giving a toal fiber length of 1–3 ft. Hemp cells are similar in dimensions, but give fiber lengths of 4–6 ft. Jute fiber bundles may be even longer, giving lengths of up to 8 ft.

The fibers of these plants are obtained from the bast by a decorticating fermentation process known as "retting".

Ramie fibers can be prepared in a highly crystalline state by simple washing procedures. The cells range from 50 to 250 mm in length and 17–64µ in diameter. Total fiber lengths are short and variable, 0.5–20 in.

Leaf fiber products included manila hemp (*Musa textilis*, Musaceae) and sisal hemp (*Agave sisalina*, Liliaceae). They are coarse and strong as indicated by their consumption in cordage and rope.

Although native celluloses have retained much of their traditional importance as fibers, recent technological advances have led to cellulose derivatives including viscose and cuprammonium regenerates, esters, and ethers. In the fiber industry, regenerated celluloses provide rayons. Viscose rayon is prepared by treating α-cellulose with 17 per cent NaOH, aging the soda cellulose under controlled conditions and converting it to sodium cellulose dithiocarbonate (xanthate) by treatment with carbon disulfide. The xanthate in alkaline solution is allowed to ripen (undergoing partial decomposition), then extruded through spinnerets into a sulfuric acid bath containing sodium and zinc salts. The xanthate is decomposed, the cellulose regenerated. In the cuprammonium process, cellulose dissolved in cuprammonium hydroxide is extruded into an aqueous regenerating bath. If the fine spinnerets are replaced with a slit-shaped orifice, the extrusion product is a film, Cellophane for example.

The viscose process, developed between 1892 and 1900, is significant in textile technology since it was the first time a fiber had been produced by chemical procedures. Although ordinary viscose rayons are lower in tensile strengths (23–30 kg/mm²) than cotton (25–80 kg/mm²), flax (50–100 kg/mm²), or ramie (90–95 kg/mm²), modern high tenacity rayons are stronger, ranging as high as 80 kg/mm².

Native celluloses have a degree of polymerization (DP) as high as 100,000 (MW $ca.$ 20,000,000), whereas regenerated cellulose gives DP 600 or more, MW 100,000 or more. When cellulose is treated with acetic acid–acetic anhydride and a catalyst (H_2SO_4,/ $ZnCl_2$), the acetylated product may be formed into bulk or sheet plastics (MW 80,000) or fibers ($MW > 100,000$). Acetate derivatives are of comparatively low crystallinity. As plastics, their tensile strength is low, 2–6 kg/mm^2 but the fibers ("acetate rayon") may range up to 75 kg/mm^2. Cellulose acetate is noted for its resistance to sunlight, humidity, weathering and biological attack.

The alkylation of celluloses to their ethers yields a series of compounds of considerable interest. Alkylation is carried out by autoclaving alkali cellulose (or cellulose+NaOH) with alkyl sulfates or chlorides. Solubility and other properties vary with the akyl group; the ratio of ether to free OH; chain length (number average molecular weight); and uniformity. In the ethyl celluloses, for example, alkali-solubility occurs when the ethoxy content lies between 5 and 15 per cent; water solubility between 20 and 30 per cent; and organic solvent solubility between 40 and 50 per cent. Various viscosity grades of water-soluble methyl celluloses have been introduced commercially for use as water soluble fibers and films, as adhesives and sizing agents and as stabilizing agents in place of natural substances such as agar. The methyl celluloses exhibit the interesting physical property of thermogel formation. When dilute (for example, 2 per cent) solutions are heated, a gel is reversibly formed at temperatures of 50–60°C. Salts and organic additives lower the gel-point.

Under dehydrating conditions, cellulose reacts with nitric acid to yield several nitroesters. The principal types and their uses include plastics, fibers (low tensile strength), explosives, etc:

Representative ester	% Nitrogen	Molecular wt.	Uses
Mononitrate	10.5–11.5	50,000	Plastics, lacquers
Dinitrate	11.5–12	150,000	Lacquers, films, fibers
Trinitrate	12–14	250,000	Explosives

Among cellulose derivatives, the nitrates have had a long history in fiber and photographic film technology, but they have been replaced by more stable products. These esters retain their importance in the form of celluloid, collodion, and gun cotton.

Among the sugar acid polymers, alginic acid, the pectic substances and agar have found application in modern technology.

Natural alginic acid is poly-1,4-anhydro-β-D-mannuronic acid, with a molecular weight of 150,000 or more. It is obtained from the cell walls of *Laminaria* in 15–40 per cent yield. As an acid, it is extracted with dilute alkali, and reprecipitated with dilute acid. The free acid has not yet been developed economically, but the salts have found several applications. Although the tensile strength of alginic acid is low (under 10 kg/mm^2) the calcium salt is comparable with the weaker celluloses and viscose rayons. Insulating foams, fireproofing fibers, and other special fibers are derived from the calcium salts. Sodium alginates are employed as thickeners, emulsifying agents, sizes, pastes and descaling agents.

Isomeric with alginic acid is pectic, or poly-1-4-α-D-galacturonic acid. As the pectic ester, it occurs abundantly in many fruits, and generally in cell walls (middle lamella). The pectins are important only as gelling and thickening agents, but they could conceivably be employed as are the alginates.

Wood: Structural

In spite of the increased modern use of metallic and synthetic strutural materials, traditional substances, such as wood, retain their importance. If the mechanical properties of wood are compared with other substances (Table 1), its relative position can be assessed.

The elastic (Young's) modulus is defined as

$$Y = \frac{F/a}{l/l_0}$$

the ratio of tensile force (F) per unit area (a) to extensions (l) produced in initial length (l_0).

This stress–strain ratio is applied within the limits of reversible deformation.

The highest elastic moduli among common materials are found in metals, alloys, and glasses. The low values among those recorded apply to various synthetic fibers and plastics. Save for wool, the natural fibers listed are appreciably more resistant to reversible deformation, than synthetics, with flax and ramie approaching the range for glasses and aluminum. The intermediate range contains the two important non-metallic structural materials, wood and concrete.

TABLE 1. MECHANICAL PROPERTIES OF WOOD AND OTHER MATERIALS

Material	Property (kg/mm^2)		
	Elastic modulus	Tensile strength	Compressive strength
Wood	900–1600	4–15	7–10
Flax, ramie (fiber)	3000–5000	50–100	10
Cotton (fiber)	600–1200	25–80	5
CaAlginate (fiber)	2000	20	—
Wool (fiber)	100–300	15–20	—
Silk (fiber)	700–1000	35–60	—
Polyethylene fiber	80	20	<2
Nylon fiber	750	80	12
Polystyrene	250–400	5	8–11
Phenol–formaldehyde	600	5–6	10–20
Concrete	2000	—	5
Glass fiber	5000–8000	125–150	60–80
Aluminum	7000	20–30	25–30 (Mg alloy)
Copper	12,500	40–45	—
Steel	20,000	100	<100

1 kg/mm^2 = *ca.* 1400 lb/in^2.

In contrast to the reversible deformation expressed by Young's modulus, tensile strength measures maximum tensile stress at the breaking point. It is, therefore, the ratio of force at breakage to initial cross-section, again F/a.

By this measure, wood is a comparatively weak material (although it approaches aluminum in some cases), whereas the stronger natural fibers approach steel and common glasses. Experimental glass fibers ten-fold stronger than ordinary steel have been produced, however. If tensile strength is divided by density, the specific strength is obtained. This corrected figure may be regarded as the maximum length of material which could be hung vertically to bear its own weight. On this basis, wood, mild steel, copper wire and many synthetics have similar values. Cotton fibers, are comparable with strong steel wire, but ramie, flax, nylon are far stronger.

Compressive (crushing) strength is a measure of the maximum stress which the material will withstand under longitudinal compression. It is expressed in the same manner as tensile strength. Wood and synthetics such as nylon or phenol-formaldehyde resins are comparable, and somewhat more resistant to compression than concrete (and ordinary brick). Glass and metals are far stronger.

The maximum stress developed at failure in the surface of standard test beams supported at their ends is the modulus of rupture, or flexural strength. This property is expressed by:

$$\text{flexural strength} = 3/2 W_m \, {}^L/bd^2,$$

Where W_m = maximum load, L = length between supports, b = width of beams, d = depth of beam. Woods, which range from 6–14 kg/mm^2, are similar or somewhat superior to flexural strength to phenol-formaldehyde resins or nylons; superior to some silicates; and from two- to three-fold lower than glass fiber and cast iron.

These comparisons taken together permit a reasonably favorable evaluation of woods for many purposes of construction. In many instances, therefore, wood may compete with other structural materials.

Other physical properties must also be considered as they affect the utility of the various structural material. The specific gravity of woods (with the exception of ebony) places them among the lighter materials for their bulk. Until the advent of expanded dolymers (e.g. polyurethane or polystyrene foams), woods were almost uniquely in this class. The thermal conductivity of wood is low when compared to most other materials, natural and synthetic. Relative to still air as a reference material, most woods have a conductivity only three-fold greater transverse to the grain and five-fold greater along the grain. Balsa wood, cork, expanded polymers, and loose fibers (glass, cotton, etc.) are among the few superior insulators.

The effect of temperature on the dimensional stability of wood is also small. The coefficient of linear expansion along the grain is generally lower than the typical values for metals, plastics and glass. Linear expansion of wood transverse to the grain exceeds iron, steel, other metals, ceramics and glasses, but is less than most synthetics. As would be expected from the other properties considered, wood is also an excellent electrical insulator.

Proper choice of structural material will, of course, depend upon aesthetic requirements and economic factors, including production costs, transportation, etc. In passing, we may wonder at the superior mechanical properties of natural fibers over wood itself, and the general basis for fiber strength.

The tensile strength of an ideal cellulose fiber made of a single continuous polymer can be calculated from the energies of (work required to break) the $C-C$, $C-H$, and $C-O$ bonds in the primary valence chain. This figure, at least, 800 kg/mm^2 far exceeds measured values for fibers. On the other hand the secondary forces holding together the molecules of a carbohydrate crystal (e.g. a sugar) give a breaking strength of only 30 kg/mm^2.

Real fibers contain discontinuous and overlapping polymer chains of imperfect and variable crystallinity. Thus, real breaking strength must depend upon secondary rather than primary valence forces. Real fibers range in tensile strength from values approximating the sugar crystal to values two to three-fold higher. Hence, the strength of real fibers derives from resistance to shearing forces which pull apart the polymer chains, by breaking the secondary bonds. Shear, like other elastic properties, is defined in terms of the stress, F/a. Thus, the displacement (strain) imparted by a given force will diminish with increasing surface as the stress itself is reduced.

It follows, therefore, that better crystallized polymer bundles will have more interchain contact surface, and will exhibit higher

rigidity or tensile strength. The effect of orientation (and consequently increased interchain force) is well illustrated in synthetics
or modified polymers which may be prepared either in bulk or
fibrous form (Table 2). The elastic modulus (reversible deformation)
is increased from two-to four-fold by orientation, and tensile
strength (breakage) may be increased by a factor of 10 or more.
The effect of orientation is evident even in the case of polyethylenes,
whose hydrocarbon chains are held together only by weak van
der Waals forces.

Wood, as a complex of cells and tissues, contains polymers
varying widely in crystallinity, together with amorphous polymers
and variously ordered small molecules. The weakness of wood
relative to the fibers themselves must then reflect the higher proportion of imperfectly aligned or weakly bonded surfaces in the
former.

Wood; Chemical

The chemical technology of wood encompasses its utility both
as a fuel and raw material. Recognizing that the combustion
properties of wood must be taken into account when wood is
used as a construction material or as a fuel, we proceed to a consideration of its chemical utilization.

The thermal decomposition of woods has long been employed to
obtain useful chemicals. Before the introduction of coking methods,
the decomposition of wood provided the charcoal required in iron
production. Two major by-products of charcoal manufacture are
methanol and acetic acid. The importance of wood distillation
as a source of alcohol has declined since the development of efficient catalystes for the direct syntheses of methanol from carbon
monoxide and hydrogen. Acetic acid derived from wood is now
largely supplanted by fermentation methods and its synthesis from
acetylene.

Although wood distillation yields only a few substances in
large quantities the diversity of minor products is impressive.
More than one hundred compounds have been identified and
include members of the following groups: carboxylic acids; aliphatic
alcohols; aliphatic aldehydes and furfurals; acetone and other

ketones; phenols and methoxyl benzenes; paraffin and olefinic hydrocarbons; benzene, polycyclic compounds and furanes; ammonia, amines and pyridines.

TABLE 2.   EFFECTS OF ORIENTATION ON MECHANICAL PROPERTIES
(IN KG/MM^2)

Property	Substance	Form	
Elastic modulus	Polyethylene	bulk	20
		fiber	80
	Polyester	bulk	200–500
		fiber	1000–2000
	Cellulose acetate	bulk	200
		fiber	350–500
	Nylon	bulk	300
		fiber	750
Tensile strength	Polyethylene	bulk	1.5–2.5
		fiber	20
	Polyester	bulk	4–9
		fiber	35–80
	Cellulose acetate	bulk	4–6
		fiber	16–75
	Nylon	bulk	7–8
		fiber	40–80

Studies comparing whole wood and cell wall components have aided in identifying the sources of some distillation products. Thus, methanol and acetic acid are derived principally from components other than cellulose or lignin; acetic acid from acetyl sources (hemicelluloses or pentosans); methanol from aromatic and nonaromatic methoxyl groups. Furane derivatives also originate in the pentosans. The phenols, methoxybenzenes, and other benzenoid compounds are derived from lignins. Cellulose contributes large amounts of water soluble glucosans, and a moderate amount of charcoal.

Acid hydrolysis, like thermal decomposition is a method for overall degradation that has long occupied the interest of wood technologists. Particular attention has been given to the saccharification process. A variety of acid treatments have been studied, ranging from high acid–low temperature processes to vigorous thermal procedures with dilute acid. The most successful saccharification processes use the technique of percolation. Wood chips or sawdust are treated with dilute sulfuric acid (0.2–1.0 per cent) at temperatures of 170–180°C. One hundred kilograms of dry coniferous wood yields about 400 kg of fermentable sugars, which in turn yield 20–24 l. of 100 per cent ethanol. The residue, principally lignin, is used as a fuel. Other methods have been developed to separate products including pentoses and furfural formed by differential hydrolysis. Residual lignocelluloses are used for plastic molding.

Alkaline hydrolysis, alkali fusion, oxidation, and hydrogenation represent other chemical treatments applied to wood and other cell wall materials. Of some interest have been the methods for production of liquid hydrocarbon fuels by pressure hydrogenation.

Wood plastics have been prepared by thermal hydrolysis and acid hydrolysis. These treatments yield residues which may be molded into figures or board. Molding compositions may be prepared by condensation of aniline, phenols, or aldehydes with sawdust and agricultural waste.

The most important chemical process in wood technology is undoubtedly pulping of wood to yield celluloses for paper manufacture and preparation of derivatives. In its essentials, chemical pulping is directed toward retention of a maximum of highly polymerized cellulose while removing substantially all of the hemicellulose and lignin. It is largely a delignification process. A multitude of pulping treatments have been developed, and have found application in the preparation of various desired grades of pulp from the specific raw materials available.

The important delignification processes include treatment with:

acid bisulfites	chlorine
alkali	nitric acid
sodium sulfite	organic solvents

The chemistry of these processes is extremely complex and cannot be treated here. It is, in fact, far from understood in detail.

In sulfite pulping, the fundamental reactions are sulfonation of the lignin, and, secondarily, acid hydrolysis of hemicelluloses. An average analysis of commercial unbleached, sulfite pulp yields about 79 per cent α-cellulose, 15 per cent cellulosons associated with cellulose, and less than 2 per cent lignin. The yield of sulfite pulp is about 50 per cent of the initial wood.

Alkaline pulping involves solubilization of lignin by release of free hydroxyl groups and hydrolysis and degradation of polyuronide hemicelluloses. Two variants are soda pulping with caustic soda, and sulfate (kraft) pulping, which is the soda process modified by addition of sodium sulfide. The unbleached pulps obtained are similar in analysis to sulfite pulps, but run to higher percentages of non-cellulosic constituents.

Bleaching with chlorine or other oxidants follows pulping and yields materials higher in α-cellulose and lower in lignins and other non-cellulosic constituents.

The spent sulfite liquors represent a major by-product of the pulping operation, and are as yet unsolved problems in industrial by-product utilization. Many millions of tons of potentially useful chemical materials representing half of the starting wood are available in sulfite and other waste liquors. The high carbohydrate content has led to the use of neutralized waste for production of alcohol and yeasts.

Alkaline hydrolysis of sulfite liquor yields vanillin and vigorous oxidation yields vanillic and protocatechuic acids. The former finds limited use as a flavoring, the latter have been proposed as acid components for polyester fibers. The possible uses for lignosulphonic acids are legion. Aside from their fuel value these substances have been proposed for: drilling mud components; fillers and extenders in furfural plastics; latex stabilizers; soil conditioners; air entrainment agents in concrete; binders; tanning; agents; asphalt modifiers; emulsifying agents; dye intermediates; etc. At present, the productive utilization of lignin sulfonates is insignificant relative to the supply, and in addition to the wastefulness itself is added the more urgent problem of stream pollution by such biologically hazardous wastes.

II. Coal and its Origins

General Characteristics

The fossil carbonaceous substances are sometimes divided into leptobiolites, coals, and bitumens. The leptobiolites are highly resistant ambers and pollens, and bitumens are derived from marine humus. Coals are metamorphosed remains of terrestrial vegetation, which may be derived from highly decayed gelatinous vegetable matter (sapropel) or vegetable debris retaining to a greater or lesser degree its biological organization (humus).

Humic–sapropelic and sapropelic coals are richer in hydrogen than humic coals, and contain relatively high proportions of resistant materials such as pollen, wax, and resin. Sapropelic types also include products of lagal origin.

Vascular plant debris is generally accepted to be the precursor of the great and familiar group of humic derivatives which include peat, lignite, bituminous and anthracite coals.

Coals may be classified mineralogically according to specific gravity, hardness, color, luster, fracture (cleavage) and texture. It will be more meaningful, from the viewpoint of origin, to examine the petrographic classificiations of coals. Botanically, coals consist of *anthraxylon* and *attritus*. The former is derived from woody tissue and consists of smooth to lustrous homogeneous black bands, stripes, and lenticular inclusions. Chemically, it consists of lignin derivatives. Attritus is dull gray to nearly black, often striped from macerated plant remains. It contains humic derivatives together with resins, lipoidal substances (fats, oils, cuticular remains, pollen coats) and minerals. These various constituents of attritus form the basis for sub-classification according to their presence and proportion. Lithological classification distinguishes *vitrain*, *clarain*, *durain*, and *fusian*. Vitrain occurs in banded or lenticular inclusions and has a woody microscopic structure. Clarain is inhomogeneous, irregular, striated or streaked whereas durain is finely granular and contains many pollen exines. Fusain is porous, dull, and charcoal-like, consisting of a cellular network often filled with pyrites, gypsum, calcite or clay. Vitrain has been further classified according to its content of jellied plant matter (*ulmain* and *collain*), cork (*suberain*), or wood (*xylain*).

The organic constituents of coals are many and varied. Indeed, one of the great problems in coal chemistry is concerned with differentiations of the fundamental structural units of coal hence with the development of fractionation procedures which are mild enough to separate the molecular species actually present without obscuring chemical alteration. The substances which can be recognized in the various coals include—

Paraffins and aromatic hydrocarbons (liquid and solid):

C_{10}, C_{11}, C_{13}, C_{17}, C_{24}, C_{26}, C_{32};

naphthalene and polynuclear hydrocarbons.

Gases (at N.T.P.):

H_2, CH_4, C_2H_6, CO, CO_2, N_2.

Salts of organic acids:

oxalates of Ca(II) and Fe(II), aluminum mellitate.

Complex organic acids:

humins, ulmins (C 50–60%, H 4.5–6.0%, N 0.5–2.0%).

Lipoidal and resinous substances:

resins, resinoles, resinolic acids (C 76.8–83.6%, H 9.7–12.9%);

waxes (C 80.3–81.6%, H 13.1–14.1%);

fats, oils (C 74–78%, H 10.3–13.4%).

Mineral constituents:

sulfur (as sulfides, sulfates, some organic S);

phosphorus (as phosphate, some organic phosphorus);

alkali metals and halogens (as Na^+, K^+, Cl^-);

alkaline earths (Ca and Mg carbonates);

silicon (as silica, silicates);

heavy metals (FeS_2, $CuFeS_2$, FeAsS, SbS_3).

The various forms of coal may be arranged according to variety and rank, a classification of importance in the consideration of the genesis of coal. *Peat* is the incipient stage in coal formation and consists of hydrated light brown material of fibrous, woody or jelly-like texture. On the average dry peats contain C 53 per cent, H 6 per cent, N 1.3 per cent, ash 10 per cent, but show considerable variability. The next rank (and stage) is *lignite* and *brown coal* which contains less water and ash, and may be somewhat richer in carbon than the lower ranks. They are progressively harder and lower in their content of water and other volatile substances. Peat may contain as much as 74 per cent non-aqueous volatile components whereas bituminous coal contains no more than 40 per cent. Semi-bituminous coals which contain no more than 20 per cent volatile matter grade into *semi-anthracites* (approximately 10 per cent volatiles or less).

The final stage in coal evolution, *anthracite* is distinguished by its hardness, low volatile content and extremely high carbon content.

Coalification

Although it has been suggested that coal might originate from sedimented vegetable matter removed from its point of origin (*drift* or allochthonous theory), the weight of evidence supports the *in situ* or *autochthonous* theory. Accordingly, coal is believed to have originated from the vegetable sediments which characterize peat bogs and freshwater swamps. Deposits which develop in poorly drained cool moist regions originate in ponds a few hundred feet to a mile in diameter.

In such areas, succession leads to the establishment of pond weeds, water lilies, rushes, algae and the formation of a floating mat. The accumulation of vegetable debris from these mats slowly fills in the center of the pond while the death of fringe growths leads to organic deposits at the bog edge. In time, peat and muck fill in the bog and allow the formation of conifer stands (tamarack, spruce) and, subsequently decidous trees such as white birch.

Although typical peat bogs are an important site for formation of lower rank coal progenitors, the freshwater swamps are the most likely source of large coal deposits. Areas such as the Dismal Swamp of North Carolina and the great Sumatra Swamp of the East Indies are probably representative in part of typical sites of coalification. In the Dismal Swamp, a peat deposit ranging from 1–20 ft in thickness (average, 7 ft) is contained in an area of 1500 square miles. It is estimated that nearly 700 million tons of peat are held in this deposit, corresponding to a coal seam 1–20 in. in thickness.

The 30 ft peat layer of the Sumatra Swamp covers an area of about 300 square miles, and would yield a 3 ft coal deposit.

The rate of peat accumulation varies widely according to type of vegetation, its growth rate, the soil, climate, and rate of decay.

In the Jura Mountains, 18–20 in. of peat have accumulated in 50 years. In the valley of the Somme, 1 in. per year accumulates and in moss areas of Denmark, 250–300 years are required for laying down of 10 ft. Under the most favorable conditions, a rate of 2 in. per year is possible. Old, compressed peat represents one--eighth the thickness of fresh deposits, hence accumulates at the rate of about 0.1 in. per year. In general 20 ft of vegetable debris or 3 ft of old peat correspond to 1 ft of bituminous coal. The transformation of vegetable debris or fresh peat into 1 ft of bituminous coal is estimated to require 300 years.

Although they must be regarded as highly empirical expressions, a number of overall equations for coal formation have been proposed on the basis of the conversion of cellulose. For example, the formation of bituminous coal from the wood of *Cordaites* has been written:

$$(C_6H_{10}O_5)_4 \rightarrow C_9H_6O + 7CH_4 + 8CO_2 + 3H_2O$$

The value of such gross equations is highly questionable.

In the transformation of vegetable matter into coal it is possible to recognize two stages—the biochemical–fermentative, and the geophysical–geochemical. In the stage I, soil bacteria in conjunction with autolytic processes macerate plant tissues, release or solubilize many constituents, and degrade or modify many of the chemical components thus made available. Bacteria are known in peat and coal, *Micrococcus* and *Streptococcus*, for example, and some forms are known to operate at depths of at least 25–30 ft in peat bogs. The level of bacterial activity must eventually become restricted to anaerobes and will eventually be checked by CO_2 and organic acids which lower the pH.

The principle events which have been definitely associated with the biochemical stage in coalification are the disaggregation of tissues and selective degradation of carbohydrates and other primary cellular components. The more resistant secondary substances such as lignin, tannins, cutin, and the like are thereby concentrated and rendered susceptible to further modification. The secondary phenolic substances, are believed to be the principle precursors of coal. Quantitatively, lignin assumes a place of greatest importance in coal formation therefore.

Experimentally, lignins can be transformed into products which resemble coals in elementary composition and certain properties. Bergius autoclaved spruce lignin which analyzed C 64.5 per cent, H 4.5 per cent, or $(C_{11}H_{10}O_4)_x$, with water at 340°C for 3 hr. This hydrothermal treatment yielded CO_2 and a product analyzing C 83.1 per cent, H 5.6 per cent. After fractionation, two coal-like substances corresponding to $(C_{10}H_{10}O)_2$ and $(C_{10}H_6O)_x$ were isolated. Bergius further claimed that "lignin" from bryophytes and lycopods was also converted to coal-like products by hydrothermal treatment. It is reasonably well established that the concentrated lignin (and tannin) residues from cell walls give rise to humic acids, and that these in turn are transformed into peats, lignites, and the higher ranks of coal.

The role of base exchange as a first stage process has been claimed by Taylor, who has described the rapid carbonization of vegetable matter at 3 m depth in the alkaline soils of Egypt. Experimentally, the rapid coalification of wood and other plant substances has been carried out by sandwiching these materials between sand and wet sodium alumino-silicates. Under such experimental conditions, methane is entrapped, CO_2 disappears, and the sand floor becomes alkaline and anaerobic.

The secondary stage is concerned with the fate of the initial residues. Among the many factors worthy of note are: extent of exposure of vegetable remains before burial; burial time; depth of burial; heat; pressure; and escape of volatile matter. A few examples will serve to illustrate the manner in which these factors influence coalification:

(a) Fusains are richer in regions which allowed dry rot of plant remains extending above water surfaces.

(b) In a general way, older coals are of higher rank than more recent deposits, presumably as a result of prolonged exposure to metamorphic processes.

(c) The relative importance of thermal processes in formation of lower rank coals is disputed and questionable, but the formation of anthracite reportedly requires temperatures of 350°C–600°C. Locally, igneous intrusions may provide heat for such transformations, but the temperature in regional deposits was probably elevated by frictional heat generated at shear surfaces. The generation of pressure is obviously associated with these regional heating effects.

Little is known with certainty but a great deal has been inferred about the details of coal formation. The general features of the coalification process and its relation to cell wall substances are reasonably well established, as should be evident even in this brief discussion.

READINGS AND REFERENCES

CONSTITUTION AND ARCHITECTURE IN THE CELL WALL

BONNER, J., *Plant Biochemistry*. Academic Press, New York (1950).
BRAUNS, F., *The Chemistry of Lignins*. Academic Press, New York (1950).
DEUEL, H. and STUTZ, E., Pectic substances and pectic enzymes. *Advances in Enzymes*. **20** 341–382 (1958).
FREY-WYSSLING, A., *Submicroscopic Morphology of Protoplasm and its Derivatives*, 2nd ed. Elsevier, New York (1953).
KERTESZ, Z., *The Pectic Substances*. Interscience, New York (1951).
MEYER, K., *Natural and Synthetic High Polymers*, 2nd ed. *High Polymers*, vol. IV. Interscience, New York (1950).
NORMAN, A. G., *The Biochemistry of Cellulose, The Polyuronides*. Lignin & C. Oxford Press (1937).
PIGMAN, W. (Editor), *The Carbohydrates*. Academic Press, New York (1957).
SMITH, F. and MONTGOMERY, R., *The Chemistry of Plant Gums and Mucilages*. Reinhold, New York (1959).
WHESTLER, R. and SANNELLA, J., Hemicelluloses. In *Proc. 4th Intl. Cong. of Biochem.* (Edited by WOLFROM, M.), vol. I. Pergamon Press, New York (1959).
WISE, L. E., *Wood Chemistry*. Reinhold, New York (1944).

CELL WALL DYNAMICS

BISHOP, C., BAYLEY, S. and SETTERFIELD, G., Chemical constitution of the primary cell walls of avena coleoptiles. *Plant Physiol.* **33**, 283–288 (1958).
BOLL. W., Ethionine inhibition and morphogenesis of excised tomato. *Plant Physiol.* **35**, 115–122 (1960).
BONNER, J., *Plant Biochemistry*. Academic Press, New York (1950).
BOROUGHS, H. and BONNER, J., Effects of indoleacetic acid on metabolic pathways. *Arch. Biochem. Biophys.* **46**, 279–290 (1953).
BRAUNS, F., *The Chemistry of Lignin*. Academic Press, New York (1950).
BROWN, S. and NEISH, A., Shikimic acid as a precursor in lignin biosynthesis. *Nature* **175**, 688–690 (1955).
BROWN, S. and NEISH, A., Studies of lignin biosynthesis using isotopic carbon—IV. Formation from some aromatic monomers. *Can. J. Biochem. and Physiol.*, **33**, 948–962 (1955).
BROWN, S. and NEISH, A., Studies of lignin biosynthesis using isotopic carbon—V. Comparative studies on different plant species. *Can. J. Biochem. and Physiol.* **34**, 769–778 (1956).

BROWN, S. and TANNER, K., Studies of lignin biosynthesis using isotopic carbon—II. Short-term experiments with $C^{14}O_2$. *Can. J. Chem.* **31**, 755–760 (1953)

CLARKE, P. and TRACEY, M., The occurrence of chitinase in some bacteria. *J. Gen. Microbiol.* **14**, 188–196 (1956).

DOUGALL, D. and SHIMBAYASHI, K., Factors affecting growth of tobacco callus tissue and its incorporation of tyrosine. *Plant Physiol.* **35**, 87–97 (1960).

FREY-WYSSLING, A., *Submicroscopic Morphology of Protoplasm and its Derivatives*, 2nd ed. Elsevier, New York (1953).

FREUDENBERG, K., Biochemische Vorgange bei der Holzbildung. In *Proc. 4th Intl. Cong. of Biochem.* vol. II, pp. 121–136 Pergamon Press, New York (1959).

FREUDENBERG, K., Biosynthesis and constitution of lignin. *Nature* **183**, 1152–1155 (1959).

FREUDENBERG, K., HARKIN, J., REICHERT, M., and FUKOZUMI, T., Die an der Verholzung beteiligten Enzyme. Die dehydrierung des Sinapinalkohols. *Chem. Ber.* **91**, 581–590 (1959).

FREUDENBERG, K. and SAKAKIKARA, A., Weitere Zwischenproduckte der Bildung des Lignins. *Liebigs Ann. Chem.* **623**, 129–137 (1959).

GARDNER, F. and COOPER, W., Effectiveness of growth substances in delaying abscission of coleus petioles. *Botan. Gaz.* **105**, 80–89 (1953).

GREATHOUSE, G., On the enzymic polysaccharide synthesis. *In Proc. 4th Intl. Congr., Biochem.*, vol. II, pp. 76–81 (1959).

HALLIWELL, G., Cellulolysis by rumen microorganisms. *J. Gen. Microbiol.* **17**, 153–165 (1957).

HESTRIN, S., Enzymic synthesis and cleavage of lavan. In *Proc. 4th Intl. Congr. of Biochem.* (Edited by WOLFROM, M.), vol. I, pp. 181–193. Pergamon Press, New York (1959).

HIGUCHI, T., Further studies on phenol oxidase related to the lignin biosynthesis. *J. Biochem.* **45**, 515–528 (1958).

HIGUCHI, T., Studies on the biosynthesis of lignin. In *Proc. 4th Intl. Congr., of Biochem.*, vol. II, pp. 161–188 (1959).

HALLIWELL, G., Cellulolytic preparations from microorganisms and from *Myrothecium verrucaria. J. Gen. Microbiol.* **17**, 166–183 (1957).

JACOBS, W., The role of auxin in differentiation of xylem around a wound. *Am. J. Botany* **39**, 301–309 (1952).

JACOBS, W. and MORROW, I., A quantitative study of xylem development in the short apex of coleus. *Am. J. Botany* **44**, 823–842 (1957).

JANSEN, E., JANG, R., ALBERSHEIN, P. and BONNER, J., Pectic metabolism of growing cell walls. *Plant Physiol.* **35**, 87–97 (1960).

JENSEN, W. and ASHTON, M., Composition of developing primary wall in onion root tip cells.—I. Quantitative analyses. *Plant Physiol.* **35**, 313–323 (1960).

KAUFMAN, W., and BAUER, K., Some studies on the mechanism of the anaerobic autolysis of *Bacillus subtilis. Proc. Soc. Gen. Microbiol.* In *J. Gen. Microbiol.* **18**, xi (1957).

KREMERS, R., The lignins. *Ann. Rev. Plant Physiol.* **10**, 185–196 (1959).

LIVINGSTON, G., *In vitro* tests of abscission agents. *Plant Physiol.* **25**, 711–721 (1950).

MACLENNAN, A., The production of capsules, hyaluronic acid, and hyaluronidase by twentyfive strains of group C., Streptococci. *J. Gen. Microbiol.* **15**, 485–491 (1956).

NEISH, A., Biosynthesis of hemiculluloses. In *Proc. 4th Intl. Congr. of Biochem.*, vol. II, pp. 82–91 Pergamon Press, New York (1959).

NORD, F. and SCHUBERT, W., Lignification. In *Proc. 4th Intl. Congr. of Biochem.*, vol. II, pp. 189–206 Pergamon Press, New York (1959).

NORMAN, A. G., *Biochemistry of Cellulose, Polyuronides*. Lignin & C. Oxford Press (1957).

ORDIN, L., ALLAND, R. and BONNER, J., Influence of auxin on cell wall metabolism. *Proc. Natl. Acad. Sci. U.S.* **41**, 1023–1029 (1955).

ORDIN, L., ALLAND, R. and BONNER, J., Methylesterification of cell wall constituents under the influence of auxin. *Plant Physiol.* **32**, 216–220 (1957).

ORDIN, L., APPLEWHITE, T. and BONNER, J., Auxin-induced water uptake by *Avena* coleoptile sections. *Plant Physiol.* **31**, 44–53 (1956).

ORDIN, L. and BONNER, J., Effect of galactose on growth and metabolism of Avena coleoptile sections. *Plant Physiol.* **32**, 212–215 (1957).

PIGMAN, W. and PLATT, D., Animal polysaccharides and glycoproteins. In *The Carbohydrates* (Edited by PIGMAN, W.). Academic Press, New York (1957).

POWELL, JOAN, Chemical changes occurring during spore germination. *Spores* (Edited by HALVORSON, H.), Am. Inst. Biol. Sci. Publ., No. 5 Washington (1957).

REEVE, R., Histological and histochemical changes in developing and ripening peaches—II. The cell walls and pectins. *Am. J. Botany* **46**, 241–248 (1959).

RICHMOND, M., The formation of lysozyme by a strain of *Bacillus subtilis*. *Proc. Soc. Gen. Microbiol.* In *J. Gen. Microbiol.* **18**, xii (1957).

SALTON, M., Studies on the bacterial cell wall—V. The action of lysozyme on the cell walls of some lysozyme sensitive bacteria. *Biochim. et Biophys. Acta* **22**, 495–506 (1956).

SALTON, M., The properties of lysozyme and its action on microorganisms. *Bacterial. Revs.* **21**, 82–100 (1957).

SCOTT, F., SCHROEDER, M. and TURRELL, F., Development of cell shape, suberization of internal surface and abscission in the lead of the valencia orange, *Citrus sinensis. Botan. Gaz.* **109**, 381–411 (1948).

SETTERFIELD, G. and BAYLEY, S., Deposition of cell walls in oat coleoptiles. *Can. J. Botany* **37**, 861–870 (1959).

SHOJI, K., ADDICOTT, F., and SWETS, W., Auxin in relation to leaf blade abscission. *Plant Physiol.* **26**, 189–191 (1951).

SIEGEL, S., Structural factors in polymerization: The matrix in aromatic biopolymer formation. In *Subcellular Particles* (Edited by HAYASHI, T.). Am. Physiol. Soc., Washington, D.C. (1959).

SIEGEL, S. The biochemistry of lignin formation. *Physiol. Plantarum* **8**, 30–32 (1955)

SIEGEL, S., The chemistry and physiology of lignin formation. *Quart. Rev. Biol.* **31**, 1–18 (1956).

SIEGEL, S., FROST, P. and PORTO, F., Effects of indoleacetic acid and other oxidation regulators on *in vitro* peroxidation and experimental conversion of eugenol to lignin. *Plant Physiol.* **35**, 163–167 (1960).

SIEGEL, S., PORTO, F. and FROST, P., Bio-regulatory activity and nitrogen function in organic compounds; antioxidant properties and their physiological significance. *Physiol. Plantarum* **12**, 721–741 (1959).

SMITH, W. K., A survey of the production of pectic enzymes by plant pathogenic and other bacteria. *J. Gen. Microbiol.* **18**, 33–41 (1958).

STAFFORD, H., Differences between lignin-like polymers formed by peroxidation of eugenol and ferulic acid in leaf sections of Phleum. *Plant Physiol.* **35**, 108–114 (1960).

STONE, J., Studies of lignin biosynthesis using isotopic carbon—I. Long-term experiments with $C^{14}O_2$. *Can. J. Chem.* **31**, 207–213 (1953).

STONE, J., BLUNDELL, M. and TANNER, K., The formation of lignin in wheat plants. *Can. J. Chem.* **29**, 734–745 (1951).

STRANGE, R., Cell wall lysis and the release of peptides in *Bacillus species. Bacteriol. Revs.* **23**, 1–7 (1959).

STRANGE, R. and DARK, F., Cell wall lytic enzymes at sporulation and spore germination in *Bacillus species. J. Gen. Microbiol.* **17**, 525–537 (1957).

THATCHER, R., Enzymes of apples and their relation to the ripening process. *J. Agr. Research* **5**, 103–116 (1915).

TRACEY, M., Celluloses. In *Biological Transformations of Starch and Celluloses* (Edited by WILLIAMS, R.). Bioch. Soc., Symp. No. 11, Cambridge University Press (1953).

WARDROP, A., The nature of surface growth in plant cells. *Australian J. Botany* **4**, 193–199 (1956).

WARDROP, A. and BLAND, D., The process of lignification in woody plants. *Proc. 4th Intl. Congr. of Biochem.*, vol. II, pp. 76–81 Pergamon Press, New York (1959).

WILSON, C. and SKOOG, F., Indoleacetic acid induced changes in uronide fractions and growth of excised tobacco pith tissue. *Physiol. Plantarum* **7** 204–211 (1954).

WU, P-H, and BYERRUM, R., Studies on the biosynthesis of pectinic methyl esters. *Plant Physiol.* **33**, 230–231 (1958).

YAGER, R., Possible role of pectic enzymes in abscission. *Plant Physiol.* **35**, 157–162 (1960).

COMPARATIVE CHEMISTRY OF INTERCELLULAR SUBSTANCES AND WALLS

ABELSON, P., Some aspects of paleobiochemistry. *Ann. N.Y. Acad. Sci.* **69**, 276–285 (1957).

ARAKI, C., Seaweed polysaccharides. In *Proc. 4th Intl. Congr. of Biochem.* (Edited by WOLFROM, M.), vol. I, pp. 15–30. Pergamon Press, New York (1959).

ASTBURY, W. T., and PRESTON, R. D., Structure of the cell wall in some species of filamentous green alga, *Cladophora. Proc. Roy. Soc. London* B. **129**, 54 (1940).

BAILEY, I. W., Evolution of the tracheary tissue of land plants. *Am. J. Botany* **40**, 4–8 (1953).

BORRADAILE, L. A. and POTTS, F. A., *The Invertebrata*, 2nd ed. MACMILLAN, New York (1935).

CIESLAR, A., The lignin content of spruce wood. *Mitt. Forstl. Versuchsw. Ost.* **23**, 30 (1897).

CUMMINS, C. S., The chemical composition of the bacterial cell wall. *Intern. Rev. Cytol.* **5**, 25–50 (1956).

EVANS, T. H. and HIBBERT, H., Bacterial polysaccharides. Sci. Report Series, No. 6., Sugar Res. Found., Inc., New York (1947).

FRAENKEL, G. and RUDALL, K. M., A study of the physical and chemical properties of the insect cuticle. *Proc. Roy. Soc. London* B. **129**, 1 (1940).

FRAENKEL, G. and RUDAL, K. M., The structure of insect cuticles. *Proc. Roy. Soc. London* B. **134**, 111. (1945).

FOGG, G. E., *The Metabolism of Algae.* John Willey, New York (1953).

FREY-WYSSLING, A., *Submicroscopic Morphology of Protoplasm and its Derivatives.* Elsevier, New York (1953).

GEISSMAN, T. R. and HINREIMER, E., Theories on the biogenesis of flavonoids. *Botan. Rev.* **18**, 77–244 (1952).

HACKMAN, R., Biochemistry of the insect cuticle. In *Proc. 4th Intl. Congr. Biochem.* (Edited by LEVENBROOK), Pergamon Press, New York (1959).

HEIDELBERGER, M., All polysaccharides are immunologically specific In *Proc. 4th Intl. Congr. Biochem.* (Edited by WOLFROM, M.), vol. I, pp. 52–66 (1959).

KENT, P. W. and WHITEHOUSE, M. W., *Biochemistry of the Amino Sugars.* Academic Press, New York (1955).

LEVIN, R. A., The cell walls of *Platymonas. J. Gen. Microbiol.* **19**, 87–90 (1959).

MANSKAJA, S. M., Zur Phylogenese des Lignins. In *Proc. 4th Intl. Cong. of Biochem.* (Edited by KRATZL, K. and BILLEK, G.), vol. II, pp. 215–226. Vienna (1958).

MANSKAJA, S. M., Lignin of various plant groups. *Akad. Nauk SSSR* **10**, 98–115 (1954).

MCLEAN, F. C. and URIST, M. R., *Bone.* University of Chicago Press, Chicago (1955).

MILLER, S. L., The formation of organic conpounds on the primitive earth. In *Modern Ideas on Spontaneous Generation. Ann. N.Y.Acad. Sci.* **69**, 260–275 (1957).

NICKERSON, W. J. and FALCONE, G., Function of protein disulfide reductase in cellular division of yeasts. In *Sulfur in Proteins.* Academic Press, New York (1959).

NORTHCOTE, W., Polysaccharides of the yeast ell wall. *Proc. Soc. Gen. Microbiol.* In *J. Gen. Microbiol.* vol. II, viii (1954).

PAECH, K., *Biochemie und Physiologie der Sekundären Pflanzenstoffe*, vol. I. Springer-Verlag, Berlin (1950).

PRESTON, R. D. and ASTBURY, W.T., The structure of the wall of the green alga *Valonia ventriosa. Proc. Roy. Soc. London* B **122**, 76 (1934).

PRESTON, R. D., NICOLI, E., REED, R. and MILLARD, A., Electron microscopic study of cellulose in the wall of *Valonia. Nature* **162**, 665 (1948).

RUDAL, K. M., The distribution of collagen and chitin in fibrous proteins and their biological significance. *Symp., Soc. Exptl., Biol.,* vol. IX. Academic Press, New York (1955).

SEN, J., Fine structure in degraded, ancient and buried wood and other fossilized plant derivatives. *Botan. Rev.* **22**, 343–374 (1956).

SIEGEL, S., The chemistry and physiology of lignin formation. *Quart. Rev. Biol.* **31**, 1–18 (1956).

SIEGEL, S., LeFEVERE, JR., B., BORCHARDT, R., Ultraviolet absorbing components of fossil and modern plants in relation to thermal alteration of lignins. *Am. J. Sci.* **256**, 48–53 (1958).

SMITH, G. M., *Cryptogamic Botany*, vol. I. *Algae and Fungi.* McGraw-Hill, New York (1938).

STRANGE, R. E., Cell wall lysis and the release of peptides in *Bacillus species. Bacteriol Rev.* **23**, 1–7 (1959).

WIGGLESWORTH, V. B., *The Physiology of Insect Metamorphosis.* Cambridge University Press, Cambridge (1954).

PROPERTIES AND USES OF CELL WALLS AND THEIR DERIVATIVES

ASTM Standard Methods for Testing Small Clear Specimens of Timber. ASTM Standard, **143**, 52, pt. 4, 720–757 (1952).

BRAUNS, F. E., *The Chemistry of Lignins.* Academic Press, New York (1950).

HOUWINK, R., *Elasticity, Plasticity and the Structure of Matter.* Cambridge University Press (1937).

KOEHLER, A., *The Properties and Uses of Wood.* McGraw-Hill, New York (1924).

MARK, H., Mechanical properties of high polymers. *Trans. Faraday Soc.* **43**, 447–462 (1947).

MCKENZIE, TAYLOR E., The decomposition of vegetable matter under soils containing calcium and sodium as replaceable bases. *Fuel* **6**, 359–367 (1927).

MCKENZIE, TAYLOR, E., Base exchange and the origin of coal. *Fuel* **5**, 195–202 (1926).

MCKENZIE, TAYLOR E., The replaceable bases in the roofs of bituminous coal seams of tertiary age. *Fuel* **7**, 129–130 (1928).

MOORE, E. S., *Coal.* 2nd, ed. John Wiley, New York (1940).

Отт, E., Spurlin, H. M. and Grafflin, M. W., (Editor), Cellulose and Cellulose Derivatives. *High Polymers* vol. V. Interscience, New York (1954).

Potonie, R., Petrographische Klassification der Bitumina. *Geol. Jahrb.* **65** (1950).

Press, J. J. (Editor), Man-made fiber progress. *Ann. N. Y. Acad. Sci.* **67**, 897–982 (1957).

Roff, W. J., *Fibres, Plastics and Rubbers.* Butterworths, London (1956).

Tomkeieff, S. I., *Coals and Bitumens.* Pergamon Press, London (1954).

Waksman, S. A., *Humus: Origin, Chemical Composition, and Importance in Nature.* London, (1938).

Wise, L. E., *Wood Chemistry.* Reinhold, New York (1944).

Wood, H., Handbook No. 72, USDA Forest Service, USGPO, Washington (1955).

INDEX OF ORGANISMS

(according to usage in text)

GENERAL SUBJECT INDEX